C. M. H. Barritt

Dip. Arch., F.R.I.B.A.

Advanced building construction
Volume 1

Illustrated by the author

Longman London and New York

Longman Group Limited
Longman House, Burnt Mill, Harlow
Essex CM20 2JE, England
Associated companies throughout the world

*Published in the United States of America
by Longman Inc., New York*

First published 1984

British Library Cataloguing in Publication Data
Barritt, C. M. H.
   Advanced building construction Volume 1
   1. Building
   I. Title
   690   TH145

   ISBN 0–582–41316–8

Set in 10/11pt Linotron 202 Times
Printed in Hong Kong by
Warren Printing Co., Ltd.

General Editor – Construction and Civil Engineering

C. R. Bassett, B.Sc., F.C.I.O.B.
*Formerly Principal Lecturer in the Department of Building and Surveying,
Guildford County College of Technology*

Books published in this sector of the series:

Building organisations and procedures   G. Forster
Construction site studies – production, administration and personnel
    G. Forster
Practical construction science   B. J. Smith
Construction surveying   G. A. Scott
Materials and structures   R. Whitlow
Construction technology Volume 1   R. Chudley
Construction technology Volume 2   R. Chudley
Construction technology Volume 3   R. Chudley
Construction technology Volume 4   R. Chudley
Maintenance and adaptation of buildings   R. Chudley
Building services and equipment Volume 1   F. Hall
Building services and equipment Volume 2   F. Hall
Building services and equipment Volume 3   F. Hall
Measurement Level 2   M. Gardner
Measurement Level 3   M. Gardner
Structural analysis Level 4   G. B. Vine
Site surveying and levelling Level 2   H. Rawlinson
Economics for the construction industry   R. C. Shutt
Design procedures Level 4   J. M. Zunde
Design technology Level 5   J. M. Zunde
Architectural design procedures   C. M. H. Barritt
Construction technology for civil engineering technicians   P. L. Monckton
Environmental science   B. J. Smith, G. M. Phillips and M. Sweeney
Concrete technology Level 4   J. G. Gunning
Construction drawing Level 1   R. Boxall
Construction processes Level 1   R. Greeno
Design of structural elements 1   A. G. Smyrell
Environmental science   B. J. Smith, G. M. Phillips and M. Sweeney
Basic accounting for builders   D. Hughes
Thermodynamics Level 3   R. Joel
Structural analysis Level 5   S. R. Mangalgiri
Hydraulics for civil engineering technicians   T. Cairney

Contents

Part II Substructure

Chapter 4 Soil testing and underpinning 36

Chapter 5 Temporary support 51

Chapter 6 Choosing a foundation 57

Chapter 7 Pile foundations 69

Chapter 8 High retaining walls 79

Chapter 18   Prefabricated stairs   172

Chapter 19   Industrial doors and shutters   183

Chapter 20   Internal joinery and adhesives   193

Part V   External works

Chapter 21   Rigid pavements   210

Preface

Advanced Building Construction volume 1 covers all the topic areas set out in the Business and Technician Education Council's unit: Building Technology IV, and its companion volume, Advanced Building Construction volume 2 covers Building Technology V. These two, taken together, also provide all the information required by students taking the new BTEC units of Building Construction A IV, B IV and V

It is of use, then, to all students at present following either a Higher Certificate or a Higher Diploma course in Building Studies for whom the subject of Building Technology IV is an essential unit and, in conjunction with its companion volume, will be of use to the same students when the new units of Building Construction A IV and B IV are introduced. However, since the study of building construction techniques is not the sole prerogative of the BTEC courses, students following other programmes of study which contain this subject will find the information contained in this book of interest and help.

Many of the topics covered are each the subject of a complete and much larger text. I have endeavoured, therefore, to concentrate on the principles behind the techniques; the reasons why a particular method has been devised and selected and the advantages of one over others in given circumstances. The illustrations are all freehand sketches, deliberately to prevent them appearing as working drawings of a particular construction, thus emphasising that they are intended to demonstrate the way a technique is employed.

This book does not represent an exhaustive study of every aspect of every subject covered – further information can be found by reference to the titles listed in the Bibliography – but I believe it covers each in sufficient detail to meet the standard set by the BTEC unit.

The year 1983 should see the publication of a completely revised set of Building Regulations but at the time of writing the information available is neither necessarily legally correct nor complete, nor is it certain that they will be published in 1983. I have, therefore, made reference to the 1976 Regulations and, where appropriate, have given the corresponding clause in the new Regulation in a note (these are signalled by a superior numeral, and are listed at the end of chapters). For convenience, I have assumed that publication will be when expected and have referred to the new legislation as the 1983 Regulations.

I have been fortunate in securing the assistance of a careful and, now, skilled manuscript reader, my son Paul, to whom I must express my grateful thanks for reading my original manuscript, looking for errors and incidentally learning a certain amount of construction technology! I am much indebted to Jane Taylor for patiently transcribing my difficult handwriting into neat typewritten copy with impressive efficiency and thanks are due to my wife, who not only helped to check the manuscript but also made the writing of this book possible by her encouragement as well as by making sure that I had the time available in which to write by helping with or by taking on many of the jobs I would otherwise have been doing.

On the technical side my thanks too to Colin Bassett for his encouragement and for correcting the manuscript and, finally, I must express my gratitude for my colleagues in the Department of Building of Southend College of Technology for the assistance they gave to me in assembling the necessary technical information.

Acknowledgements

Our Figs. 1.2, 1.3 and 1.4 reproduced from the CI/SfB Construction Indexing Manual with the kind permission of the publishers RIBA Publications Ltd.

Part I

General

Chapter 1

Information retrieval

1.1   Classification and coding

'The world is composed of two classes of people – those who believe that the world is composed of two classes of people, and those who do not.' This cyclic classification is particularly useless but serves to show that for our fundamental psychological habit of classifying things to be of any benefit it must produce a list which satisfies the users' requirements. This is especially true in building where the end result is only obtained by the assemblage of a very large number of small parts, the control of which is greatly assisted by relevant classification of the information concerning those parts.

Correct classification relates similar and separates different lists of information by the use of facets – sets of classes which reflect one principle. Figure 1.1 illustrates how a random but alphabetical list can be classified by principles and then further sub-divided by primary and secondary function. Classifying information is an essential exercise if order is to be maintained but if that information is to be picked out from the collection with ease it must be divided in such a way that it meets the needs of the users, is labelled for easy and clear reference and is arranged to be quickly and easily found or retrieved.

Labelling for easy reference is only necessary if the words normally used to describe or classify that information are inconveniently long or numerous or if they could be ambiguous. Too often reference codes and abbreviations are used because someone is too lazy to use descriptive words whereas, correctly employed, a code can be a symbol which aids

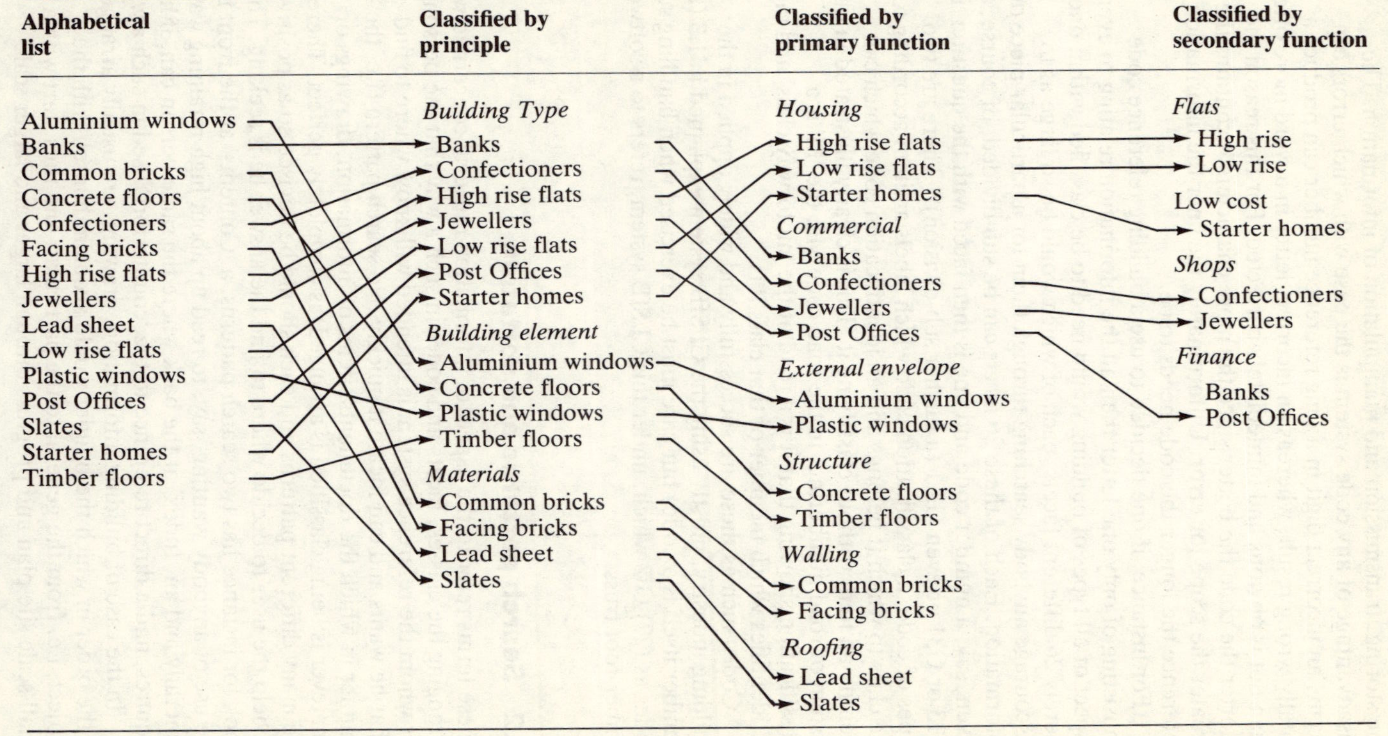

Fig. 1.1   Classifying information

the storing, transmission and manipulation of information. The disadvantage of any code system is the ease with which errors can occur, an incorrect digit in a long reference number can produce a totally wrong result, whereas an incorrect letter in a word more often than not is obvious and immediately corrected. For this reason the shorter the code the better since the more characters it contains the greater the scope for error. Unfortunately the finer the information reference the longer the code needs to be.

For instance, if one decided to use a building reference code consisting of only one letter then all the information relating to every aspect of all types of building would need to be classified within one of merely 26 different facets, each of which would be so large as to discourage anybody searching through them for a particular piece of information. Each of these 26 facets can be sub-divided, of course, and given, say, a number code and one is then faced with the question 'is it G13 or G31 or even G3.1 (a further sub-division)?' There are two types of code, a classification code which labels an item according to the class to which it belongs and an identification code which can identify an item without classifying it. This is clearly illustrated in our system of road numbering: in the road code M6, the 'M' is a classification showing that it is a road of the 'motorway class' and the '6' identifies which member of that class it is.

Codes such as these are successfully and widely applied in the building industry, mostly using the CI/SfB system explained in 1.3 (this number itself is a code) but care must be exercised when handling such codes as (27) Gi2 which, under this CI/SfB system, refers to a softwood timber roof truss.

1.2 Search patterns and sortation

These terms refer to the aspect of information retrieval concerned with arranging the classes, facets and codes in such a way that the person for whom the reference system is intended will know where to find what he wants and can locate it quickly. No search pattern (i.e. the manner in which the documents, items of information, drawings or whatever is being classified is arranged) is completely perfect. There are many different patterns and each should be devised so as positively to help the user to decide where to find the answer he is seeking. This book for instance has two search patterns, a 'Contents' at the front for the use of anybody wanting, say, to read up about high retaining walls generally and an 'Index' at the back where the same person can, if he requires, find a direct reference to the 'counterfort' used in such a wall.

In the case of building information, drawings are usually arranged in the order in which the builder would acquaint himself with the project, i.e. from the general to the particular, and thus start with a small-scale site plan and progress through larger and larger scale drawings of increasingly fine detail culminating in drawings showing aspects of the proposals to full size.

1.3 The CI/SfB system

The CI/SfB system is a classification of construction information co-ordinated in such a way as to make it more readily available to everybody in the industry.

The SfB system was originally designed in Sweden between 1946 and 1950 to provide a common classification structure for the three reference books, *Building Products Data Sheets*, *National Specification*, and *Building Price Book*. The letters SfB stand for Samarbetskomitten for Byggnadsfragor (the Technical Secretariat of the Co-ordinating Committee for Building). In 1959 it was adapted as the recommended international classification system by the International Council for Building Research Studies and Documentation and introduced into this country between 1961 and 1964 by the RIBA who became the UK licence holders for the system. RIBA Services Ltd has now taken over the responsibility for development and administration of the system. Following its introduction, development of the system took place by the RIBA and, in 1968, the original work was superseded by a second manual entitled *CI/SfB Construction Indexing Manual*, the letters CI standing for Construction Industry. Shortly after this was published the then Ministry of Public Buildings and Works was looking into data classification and produced the *Construction Industry Thesauras* which helps by standardising terminology and the British Standards Institution published BS 4940: *Recommendations for the Presentation of Technical Information* about products and services in the construction industry.

A second edition of the CI/SfB *Construction Indexing Manual* was published in 1976, updating the 1968 version and making certain fundamental alterations to increase the cover of the theoretical aspects of building.

1.4 CI/SfB tables and codes

The classification system is based on five tables: Table 0 deals with all subjects concerned with the physical environment – planning, offices, churches, housing, etc.; Table 1 covers building elements – walls, floors, roof, windows, doors, etc.; Table 2 deals with constructions – blockwork, flexible sheet work, thick coating work, etc.; Table 3 classifies materials – natural stone, glass, clay, etc.; and Table 4 lists activities and requirements – communications, transport plant, working factors, etc. The main tables in detail are shown in Figs 1.2 to 1.5.

In Table 0 (Fig. 1.2) the ten single numeral references 0–9 are

Physical environment

Planning areas

0 *Planning areas*
01 Extra terrestrial areas
02 International, national scale planning areas
03 Regional, sub-regional scale planning areas
04
05 Rural, urban planning areas
06 Land use planning areas
07
08 Other planning areas
09 Common areas relevant to planning

Facilities

1 *Utilities, civil engineering facilities*
11 Rail transport
12 Road transport
13 Water transport
14 Air transport, other transport
15 Communications
16 Power supply, mineral supply
17 Water supply, waste disposal
18 Other

2 *Industrial facilities*
21–25
26 Agricultural
27 Manufacturing
28 Other

3 *Administrative, commercial, protective service facilities*
31 Official administration, law courts
32 Offices
33 Commercial
34 Trading, shops
35–36
37 Protective services
38 Other

4 *Health, welfare facilities*
41 Hospitals
42 Other medical
43
44 Welfare, homes
45
46 Animal welfare
47
48 Other

5 *Recreational facilities*
51 Refreshment
52 Entertainment
53 Social recreation, clubs
54 Aquatic sports
55
56 Sports
57
58 Other

6 *Religious facilities*
61 Religious centres
62 Cathedrals
63 Churches, chapels
64 Mission halls, meeting houses
65 Temples, mosques, synagogues
66 Convents
67 Funerary, shrines
68 Other

7 *Educational, scientific, information facilities*
71 Schools
72 Universities, colleges
73 Scientific
74
75 Exhibition, display
76 Information, libraries
77
78 Other

8 *Residential facilities*
81 Housing
82 One-off housing units, houses
83
84 Special housing
85 Communal residential
86 Historical residential
87 Temporary, mobile residential
88 Other

9 *Common facilities, other facilities*
91 Circulation
92 Rest, work
93 Culinary
94 Sanitary, hygiene
95 Cleaning, maintenance
96 Storage
97 Processing, plant, control
98 Other; buildings other than by function
99 Parts of facilities; other aspects of the physical environment; architecture, landscape

Fig. 1.2 CI/SfB Table 0

Elements

Substructure

(1–) *Ground, substructure*
(10)
(11) Ground
(12)
(13) Floor beds
(14)–(15)
(16) Retaining walls, foundations
(17) Pile foundations
(18) Other substructure elements
(19) Parts of elements (11) to (18)
      Cost summary

Structure

(2–) *Primary elements, carcass*
(20)
(21) Walls, external walls
(22) Internal walls, partitions
(23) Floors, galleries
(24) Stairs, ramps
(25)–(26)
(27) Roofs
(28) Building frames, other primary
      elements
(29) Parts of elements (21) to (28)
      Cost summary

(3–) *Secondary elements, completion*
      if described separately from (2–)
(30)
(31) Secondary elements to external
      walls; external doors, windows
(32) Secondary elements to internal
      walls; internal doors
(33) Secondary elements to floors
(34) Secondary elements to stairs
(35) Suspended ceilings
(36)
(37) Secondary elements to roofs;
      rooflights, etc.
(38) Other secondary elements
(39) Parts of elements (31) to (38)
      Cost summary

(4–) *Finishes*
      if described separately
(40)
(41) Wall finishes, external
(42) Wall finishes, internal
(43) Floor finishes
(44) Stair finishes
(45) Ceiling finishes
(46)
(47) Roof finishes
(48) Other finishes to structure
(49) Parts of elements (41) to (48)
      Cost of summary

Services

(5–) *Services, mainly piped and*
      ducted

(50)–(51)
(52) Waste disposal, drainage
(53) Liquids supply
(54) Gases supply
(55) Space cooling
(56) Space heating
(57) Air conditioning, ventilation
(58) Other piped, ducted services
(59) Parts of elements (51) to (58)
      Cost summary

(6–) *Services, mainly electrical*
(60)
(61) Electrical supply
(62) Power
(63) Lighting
(64) Communications
(65)
(66) Transport
(67)
(68) Security, control, other
      services
(69) Parts of elements (61) to (68)
      Cost summary

Fittings

(7–) Fittings
(70)
(71) Circulation fittings
(72) Rest, work fittings
(73) Culinary fittings
(74) Sanitary, hygiene fittings
(75) Cleaning, maintenance fittings
(76) Storage, screening, fittings
(77) Special activity fittings
(78) Other fittings
(79) Parts of elements (71) to (78)
      Cost summary

(8–) *Loose furniture, equipment*
(80)
(81) Circulation loose equipment
(82) Rest, work loose equipment
(83) Culinary loose equipment
(84) Sanitary, hygiene loose
      equipment
(85) Cleaning, maintenance loose
      equipment
(86) Storage, screening loose
      equipment
(87) Special activity loose
      equipment
(88) Other loose equipment
(89) Parts of elements (81) to (88)
      Cost summary

External, other elements

(9–) *External, other elements*
(90) External works
(98) Other elements
(99) Parts of elements
      Cost summary

Fig. 1.3 CI/SfB Table 1

Constructions

A	*Constructions, forms	P	Thick coating work
B	*	Q	
C	*	R	Rigid sheet work; rigid sheets
D	*	S	Rigid tile work; rigid tiles
E	Cast in situ work	T	Flexible sheet work; flexible sheets
F	Block work; blocks	U	
G	Large block, panel work; large blocks, panels	V	Film coating and impregnation work
H	Section work; sections	W	Planting work; plants, seeds
I	Pipe work; pipes	X	Work with components; components
J	Wire work, mesh work; wires, meshes	Y	Formless work; Products
K	Quilt work; quilts	Z	Joints
L	Flexible sheet work (proofing); flexible sheets (proofing)		*Used for special purposes eg specification*
M	Malleable sheet work; malleable sheets		
N	Rigid sheet overlap work; rigid sheets for overlapping		

Materials

a	*Materials	q	Lime and cement binders, mortars, concretes
b	*	r	Clay, gypsum, magnesia and plastic binders, mortars
c	*	s	Bituminous materials
d	*		

Formed materials e/o

Functional materials t/w

e	Natural stone	t	Fixing and jointing materials
f	Precast with binder	u	Protective and process/property modifying materials
g	Clay (dried, fired)	v	Paints
h	Metal	w	Ancillary materials
i	Wood		
j	Vegetable and animal materials	x	
k		y	Composite materials
m	Inorganic fibres	z	Substances
n	Rubbers, plastics etc		
o	Glass		*Used for special purposes eg resource scheduling by computer*

Formless materials p/s

p   Aggregates, loose fills

Fig. 1.4   CI/SfB Tables 2 and 3

Activities, requirements

Activities, aids	**Requirements, properties**
(A) Administration and management activities, aids	(E/G) Description [2]
(Af) Administration, organization	(E) Composition [2.01/2.03]
(Ag) Communications	(F) Shape, size [2.04/2.06]
(Ah) Preparation of documentation	(G) Appearance [2.07]
(Ai) Public relations, publicity	
(Aj) Controls, procedures	(H) Context, environment [3]
(Ak) Organizations	
(Am) Personnel, roles	(J/T) Performance factors [4/5]
(An) Education	(J) Mechanics [4.01]
(Ao) Research, development	(K) Fire, explosion [4.02]
(Ap) Standardization, rationalization	(L) Matter [4.03/4.06]
(Aq) Testing, evaluating	(M) Heat, cold [4.07]
	(N) Light, dark [4.08]
(A1) Organizing offices, projects	(P) Sound, quiet [4.09]
(A2) Financing, accounting	(Q) Electricity, magnetism, radiation [4.10]
(A3) Designing, physical planning	(R) Energy [4.11]
(A4) Cost planning, cost control, tenders, contracts	Side effects, compatibility [4.12/4.13]
(A5) Production planning, progress control	Durability [4.14]
(A6) Buying, delivery	(S)
(A7) Inspection, quality control	(T) Application [5]
(A8) Handing over, feedback appraisal	
(A9) Other activities; arbitration, insurance	(U) Users, resources
	(V) Working factors [6]
	(W) Operation, maintenance factors [7]
(B) Construction plant, tools	
(B1) Protection plant	(X) Change, movement, stability factors
(B2) Temporary (non protective) works	
(B3) Transport plant	
(B4) Manufacture, screening, storage plant	(Y) Economic, commercial factors [18/10]
(B5) Treatment plant	
(B6) Placing, pavement, compaction plant	(Z) Pheripheral subjects; forms of presentation: time; space [11]
(B7) Hand tools	
(B8) Ancillary plant	Note: Codes in square brackets above
(B9) Other construction plant, tools	are from *CIB Master Lists for structuring documents relating to buildings, building elements, components, materials and services* (CIB Report No 18, 1972).
(C) *	
	Used for special purposes
(D) Construction operations	
(D1) Protecting	
(D2) Clearing, preparing	
(D3) Transporting, lifting	
(D4) Forming: cutting, shaping, fitting	
(D5) Treatment: drilling, boring	
(D6) Placing: laying, applying	
(D7) Making good, repairing	
(D8) Cleaning up	
(D9) Other construction operations	

Fig. 1.5 CI/SfB Table 4

major headings and are divided into sub-headings indicated by the double numeral references. A further division into sub-sub-headings can be made and would be indicated by a third numeral. Details of these sub-sub-headings can be found in the *Construction Indexing Manual*. The degree of precision required will dictate the complexity of the code. Thus, if the subject is, say, commercial buildings generally the code 3 would be adequate, whereas if the subject was shops generally then the more specific reference of 34 would be needed. Taking it one stage further, if the goods to be sold in the shops are shoes, the code 346 would be used.

In Table 1 (Fig. 1.3), as in Table 0, the main headings are indicated by a single numeral but in this case they are followed by a dash and enclosed in brackets. Similarly they are sub-divided, as shown, by adding a second numeral and can be further sub-divided by the addition of a third numeral following a decimal point. Thus electrical services generally would be (6–), security fittings (68) and emergency lighting fittings associated with security (68.6).

Tables 2 and 3 (Fig. 1.4) are concerned with the form a product takes and the materials from which it is made. The codes in this case are letters, capitals in Table 2 and lower case in Table 3 and are combined, as appropriate, to relate to a particular product. Thus any product which is a section, i.e. has a particular cross-sectional shape but no specific length (such as steel beams) would be coded H, and in this case, because the nature of the material is known, the code letter h from Table 3 would be added to show that they were metal sections. This, however, does not go far enough because they are not just metal but steel sections and therefore another sub-division can be introduced by adding the appropriate number, so that if these were mild steel beams the code would be Hh2, but if the subject was, say, a section extruded from a steel alloy such as stainless steel the code would be Hh3.

Table 4 (Fig. 1.5) deals with the activities to be found in most industries (organisation, training, communication, testing, etc.) including those peculiar to the building industry (construction, making good, cleaning up, etc.) and what is described as requirements – the subjects arising from the demands that a building should be stable, comfortable, durable and in accordance with all the performance standards expected.

As can be seen, the main codes comprise capital letters enclosed in brackets, Section (A) is further sub-divided by either a lower-case letter or a numeral within the bracket and sub-sub-divided by the addition of a lower-case letter to give, for instance, a reference (A7q) for any information concerned with testing for quality.

The other categories (B) to (Z) of the table can be similarly sub-divided and sub-sub-divided by the addition of a number and a lower case letter:

(B) Construction plant.

(B5) Treatment plant (in connection with construction).
(B5e) Welding equipment (with which to carry out the treatment in connection with construction).

1.5 Combined codes

To extend the facility for the precise classification and retrieval of information still further codes derived from any or all of the tables can be combined.

For example, referring to the codes given in the last section, if the shoe shops are to have emergency lighting fittings the combined code for this subject would be 346 (68.6). Similarly, if the information required concerns the testing of the quality of stainless steel it would be found in a file marked Hh3 (A7q); stainless steel emergency lighting fittings would, therefore, carry the reference code of (68.6) Hh3. Taking this to its ultimate refinement by combining all the codes results in the reference 346 (68.6) Hh3 (A7q) which would lead to the very narrow area of information regarding the quality testing of stainless steel emergency lighting fittings for shoe shops.

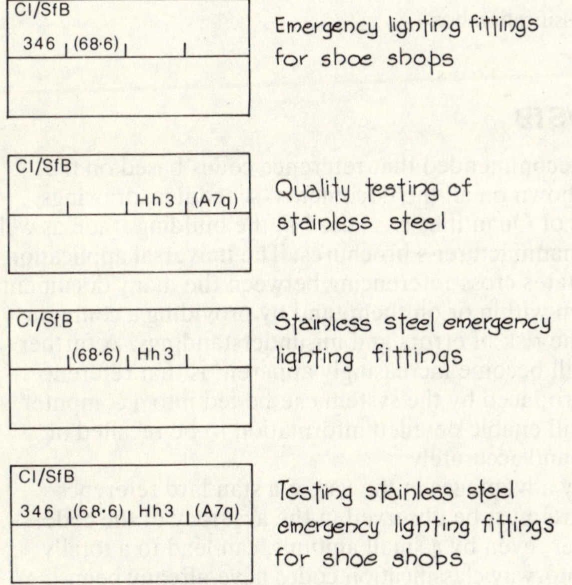

Fig. 1.6 Combined codes

1.6 CI/SfB classification box

It is now common practice to print the appropriate reference code on the front cover of building-trade literature. This is shown in the top right-hand corner of the document, placed inside a standard box as shown in Fig 1.7. As can be seen, the box is divided into an upper and lower half, the CI/SfB reference being placed in the upper and any other reference (often a Universal Decimal Classification reference) placed in the bottom half.

Universal Decimal Classification (UDC) is a completely numeric system used by libraries to classify all information. The CI/SfB half of the box is further sub-divided into four sections and codes derived from Table 0 are placed in the left-hand space, from Table 1 in the next, from Table 2/3 in the next and from Table 4 in the right-hand space. Figure 1.7 shows how all the combined codes in Section 1.5 would appear.

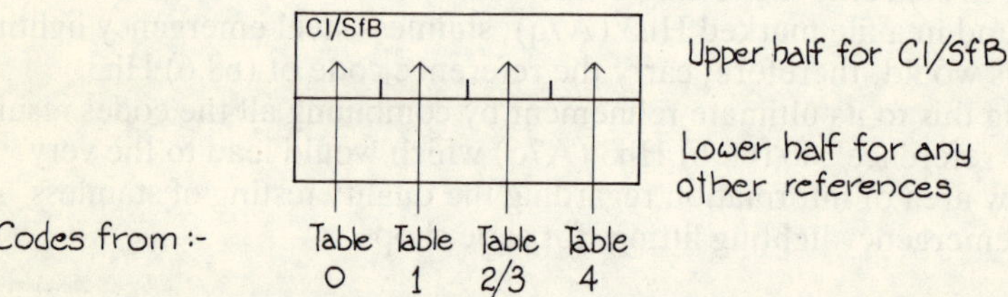

Fig. 1.7   CI/SfB Classification box

1.7   Use of CI/SfB

It is intended and recommended that reference codes based on the CI/SfB tables are shown on all the documents, schedules, drawings, specifications, Bills of Quantities etc., used in the building trade as well as on the front of manufacturer's brochures. The universal application of the system facilitates cross-referencing between the many documents and the items shown within or on them, and by providing a common language reduces the risk of errors and misunderstandings. A further advantage which will become increasingly apparent is that reference codes of the type produced by the system can be fed into a computer quite readily and will enable detailed information to be recalled or transmitted swiftly and accurately.

There are many advantages in the use of a standard reference system but great care must be observed in the accuracy of the code. Error in one number, even by a small amount, can lead to a totally different result: motorway classification codes have already been quoted at the start of this chapter, and it is not difficult to appreciate the confusion which could arise by a simple error resulting in somebody driving for miles down the M5 instead of the M6 looking for Coventry!

Chapter 2

Building standards and control

2.1 The development of the Building Regulations

All buildings, apart from certain specified types and building work in London, are required to be designed and constructed in accordance with the national standards laid down in the Building Regulations. This has not always been the case. The standards of building in the past have been set by each local authority and incorporated in their building byelaws.

The earliest legislation on this subject in this country was in London and followed the Great Fire of 1666. This was a building code intended, not unnaturally, to prevent the outbreak and spread of fire and enforced by 'discreet men, knowledgeable in building'. These 'discreet men' were the forerunners of the district surveyors who are still the people responsible for ensuring compliance with the London Building Acts.

This example was not followed anywhere else and has left London with a unique system of control. Elsewhere, a few towns and cities had local Acts which were intended to control building with respect to fire hazard and sanitation, but generally very little formal control existed.

In 1875 the Government passed a Public Health Act which gave all local authorities the power to control building construction standards in relation to safety, fire prevention, health and sanitation through local byelaws written by the authority. Guide-lines on how this should be done were laid down by the Government in a set of Model Byelaws which the local authorities could adopt with local variations. The primary concern of this early legislation was housing and the rules were a simple statement of what had to be done. The need for greater flexibility became apparent as building became more complex and the introduction of steel and reinforced concrete made use of a mathematical analysis of structure. This was recognised in the amendments to the Building Byelaws introduced through the Public Health Act of 1936.

By 1953 the rapid development in building techniques which followed the Second World War had led to a greater use of the analytical approach rather than rule-of-thumb methods, and a much wider choice of constructional systems and materials and consequently great changes were introduced. The most significant was the appearance of purely functional requirements in the byelaws, that is, a regulation stating what a system of construction must achieve rather than what it must be. Working with these rules meant that the designer was free to choose any system he considered appropriate but had to prove that it would comply with the functional requirement. To assist architects and builders in this, the simpler, more common methods of building were incorporated in 'deemed-to-satisfy' provisions which, if followed, would satisfy the mandatory functional requirements.

Not only were building methods becoming more complex but there were an increasing number of firms of architects and of builders operating over larger areas, some nationally, and for them the problems of dealing with slightly differing sets of byelaws from the 1400 local authorities then in existence presented great difficulties.

To improve this state of affairs, the Public Health Act of 1961 enabled the Minister to make national Building Regulations. The first of these came into force in February 1966, followed by a series of seven amendments which were all incorporated in the Metric Regulations of 1972. Further amendments followed and the Regulations were again revised in 1976. Subsequent amendments have again been made.

In 1981 the paper *The Future of Building Control* was published, followed in 1982 by *The Form of The Building Regulations* which dealt with the way in which the presentation, expression and management of the regulations might be improved. This led to the production of consultation documents which, at the time of writing, are still under consideration.

These consultation documents consist of three sets of papers titled *Regulations for Design and Construction*, *Appendix of Samples of Approved Documents* and *Manual to the Building Regulations*. The first of these are the requirements which must be met in building, expressed in simple functional terms based substantially on the 1976 Regulations, the second contains all the 'deemed-to-satisfy' provisions in the 1976 Regulations as well as a number of other sources of information such as Codes of Practice which set out the good building practice necessary to meet the Regulation requirements and the third set of papers provides a link between the two in the form of an index or source book.

It is intended to bring these proposals into effect in 1983 and, having effected a revision to the form of the Regulations, it is proposed to continue with revising and updating the legislation in an endeavour to keep abreast of new developments and gradually to raise building standards further.

Up to 1974 the control of building standards was derived from provisions in the Public Health Acts, but in that year the Health and Safety At Work (Etc.) Act was passed, Part III of which creates provisions for the Secretary of State to make regulations on every element in every type of building. The intention is to consolidate all building control completely in one piece of legislation (at present there are many other Acts of Parliament in addition to the Public Health Acts which control building to some extent) and to bring in educational and Crown buildings which are exempt from the Regulations. It is also proposed to bring the London Building Acts and the Scottish Building Standards into line. The process started with the Building (First Amendment) Regulation of 1978 which was the first piece of building legislation made under the Health and Safety At Work (Etc.) Act and which was concerned, in the main, with thermal insulation in buildings other than dwellings.

2.2   The powers of a local authority

Although the Building Regulations are a national standard set by Central Government the duty of interpreting and enforcing them is vested in the local district councils on whom a Notice of Intention to Build must be served before any building work can start. This Notice and the requisite drawings which accompany it are taken to be an application for Building Regulations Consent and demonstrate the applicant's intention to comply with the law.

The local authority's appointed officer will, on behalf of the authority, examine the proposals, call for further information as considered necessary and, when satisfied, will issue a Consent notice. (This has nothing to do with Town Planning permission.) Failure to satisfy the Building Control Officer that the proposals will comply with the Building Regulations will result in a refusal notice which means building work cannot commence.

A Consent notice will be accompanied by a set of blank notice forms to be completed by the builder at appropriate stages and sent to the local authority stating that the work is ready for inspection. Should any of these not be sent, the local authority has the power to require any work to be opened up or pulled down which prevents the authority from checking whether any of the Regulations have been contravened.

Not only are the local authorities given the necessary powers to enforce the Regulations, but they are also empowered to relax, or dispense entirely, the requirements of many of the Regulations should a particular requirement be shown to be unreasonable in a specific case. Unreasonableness is the only grounds for such a relaxation or dispensation. Typical of this type of consideration would be an extension to a two-storey property. Clearly the ground and first floor of the extension should be at the same levels as those existing but in making them so the builder may contravene Regulation K8 which requires that a room height is to be not less than 2.3 m.[1] In such a situation the developer would apply for and probably obtain a relaxation of this particular requirement. The local authority cannot relax or dispense with any of the requirements of Part A of the Regulations which deals with the interpretation of the Regulations. A number of general provisions such as the giving of Notice of Intention to Build, Part D, which deals with structural stability, or some of the requirements of Part E, which covers structural fire precautions, can only be dealt with by the Secretary of State.[2]

2.3 Types of Building Regulation clause[3]

Following the principle of stating functional requirements established in the Model Byelaws of 1953, the current Regulations are all written in terms of what the particular element must achieve with regard to specific requirements, functional requirements or performance requirements, as illustrated by the following requirements for external walls:

Specific requirement:
Regulation E7 (3) Any external wall which is situated within a distance of 1 m from any point on the relevant boundary . . . shall:
   (a) be constructed wholly of non-combustible materials . . .

Functional requirement:
Regulation E7 (3) Any external wall which is situated within a distance of 1 m from any point on the relevant boundary . . . shall:
   (b) be so constructed that any fire resistance required by these Regulations is attained by the non-combustible part alone.

Performance requirement:
Regulation E7 (5) Any part of an external wall of a building of purpose group VII . . . shall comply with the following provisions . . .
   (b) any such part . . . shall . . . if the outside were to be exposed to fire . . . resist the action of fire . . .

Thus, in the first example, the Regulation specified the materials which the wall must be built; in the second example it stated what the structure was expected to attain in its function as a wall and in the third example how it was required to perform when subjected to certain tests.[4]

These are the mandatory requirements; that is, these are the provisions of the Regulations with which all designers and builders must comply; but again, following the 1953 precedent, many of the Regulations are 'deemed-to-satisfy' provisions and describe constructional methods and materials which are accepted as complying with the mandatory requirements. These do not have to be followed but by doing so the designer or builder is relieved of the necessity to prove compliance with the relevant mandatory requirement.

Referring back to the examples given above, an example of a 'deemed-to-satisfy' provision in connection with external walls is set out in:

Regulation E1 (5) . . . an element of structure shall be deemed to have the requisite fire resistance if . . .
(b) in the case of a wall . . . to which Schedule 8 relates, it is constructed in accordance with one of the specifications set out in that Schedule.

Schedule 8 is one of twelve schedules which, in this case, precisely describes a range of materials, stating its specification and thickness, and methods of construction which will produce structures giving from one-half hour to four hours fire resistance. To differentiate them from mandatory requirements, 'deemed-to-satisfy' provisions are printed in italics.[5]

2.4 British Standard Specifications

Recommended product standards in this country are the responsibility of the British Standards Institution and are published as British Standard Specifications. The adoption of these Standards by the industries concerned lead to a better product of a reliable quality and, because the number of variants is limited, generally a lower cost.

The reduction of variants within a particular type of product is clearly demonstrated by the first Standard Specification written which was concerned with tramway rails and reduced the various sizes and shapes of rail produced by different steel mills from 75 to 5 and the sizes of structural steel sections from 175 to 113 (BS No. 4). This initial work was done by the Engineering Standards Committee, the first meeting of which was in 1901, chaired by Sir John Wolfe Barry and attended by representatives from the Institute of Civil Engineers, the Institute of Mechanical Engineers, the Institute of Naval Architects and the Iron and Steel Federation. In 1918 the Committee became the British Engineering Standards Association and eventually, in 1931, the British Standards Institution.

Standard Specifications are published on a very wide range of subjects not the least being products intended for the building industry which are summarised in the *Building Handbook*.

Compliance with British Standards is, unless enforced by other legislation, purely voluntary but goods marked with the familiar kitemark (Fig. 2.1) of the British Standards Institution carry with them an assurance of quality. Not all British Standards are concerned with quality, however. Some lay down standard sizes and yet others define test procedures. In quoting, or requiring adherence to, a particular standard the number quoted should be checked and the Standard itself read in full, rather than the summary mentioned above, to ensure that that particular Standard does cover the subject and the aspect required. For instance, BS 644 deals with wood windows but methods of testing windows are in BS 5368 and BS 4315. However, if the subject is the timber of which the window is made BS 1186 is required, and so on.

Fig. 2.1 British Standards Institution registered certification trade mark (kite mark)

Although these Standards are not intended to be mandatory they can and are made so by being incorporated into legal provisions such as the Building Regulations. For example:

Regulation L14 (1) Subject to the provisions of paragraph (5), any chimney serving a Class II appliance ... shall be ...
(a) lined with ...
(iii) glazed, rebated or socketed clay flue linings complying with BS 1181 1971 ... [6]

2.5 British Standard Codes of Practice

In most cases, it is not enough just to ensure that the materials are of a good standard: the way they are used and assembled must also be to a good standard if the finished work is to achieve the desired results. For this purpose, these Codes of Practice have been produced. They take the form of recommendations which, if followed, produce sound practice but compliance with these recommendations does not remove the necessity to satisfy the requirements of any legal enactments.

As with the Standard Specifications, many of the Codes of Practice have been included in other legislation, such as the Building Regulations, and thereby made mandatory, as in the following:

> Regulation D2 Calculation of loading
> (2)(a) dead loads shall be calculated in accordance with CP3 : Chapter V : Part 1 : 1967.[7]

The origination of the Codes of Practice is not as old as the Standard Specifications. It was in September 1942 that the Council for Codes of Practice for Building was formed under the aegis of the then Ministry of Works. This Council comprised nominees from fifteen professional and scientific institutions concerned with building plus representatives of the British Standards Institution and the Government who then embarked on their appointed task of preparing Codes of Practice for buildings, their construction and their engineering services.

The Codes are issued by the British Standards Institution on behalf of the Council.

Notes

1. Proposed to be A.09 in the 1983 Regulations and the specific requirement of 2.3 m probably will be transferred to Approved Document AD.A.02.
2. The subject of relaxations is included in the 1983 Regulations but not defined in the draft.
3. The proposed 1983 Regulations are all written in functional terms. (See page 14 above)
4. The requirements of Regulation E7 are to be incorporated in the proposed Regulation B.12 of the 1983 Regulations.
5. The requirements of Regulation E1 are proposed to be incorporated in B15 of the 1983 Regulations.
6. Proposed to be B.20 in the 1983 Regulations in which the reference to BS 1181 is to be omitted.
7. It is proposed that this Regulation shall become B.02 in the 1983 Regulations which will require the 'ability to sustain safely the combined dead, wind and imposed load . . .' (or some similar wording) and CP3 will become an Approved Document. (See page 14 above)

Chapter 3

Site control of materials and workmanship

3.1   The need for inspection

In the long term, every part of every element of every building is
tested, unwittingly, by the occupant simply by using it. Every aspect,
from the ability of the foundations to convey the building loads to the
ground in safety to the capacity of a door handle to withstand constant
use, will be required to meet the specified performance for an
appropriate period of time – the foundations would, naturally, be
expected to last longer than the door handle.

The breakdown of any part can be attributed either to a bad
decision by the design team or to the finished product not being in
accordance with that design decision as specified. It is to cover this
latter possibility that careful inspection is required throughout the
construction process.

Not only must the inspection be carried out with care but it should
also be done at the right time. The cost of replacing damaged or
defective goods escalates with the delay in their discovery and it can be
shown that the cost of stripping out, replacing and making good work
which is found to be defective after completion can be as much as three
times the original cost and can represent the difference between a
profit and a loss for the builder. It is, therefore, most important that
all materials are thoroughly checked and inspected when they first
arrive on site.

Workmanship cannot be inspected in this way and must receive
constant supervision to ensure that one stage is correct before

proceeding to the next. For instance, it is not difficult to move a trench before concrete is placed in it but it is a very different matter to change the position of the concrete foundation once set; and therefore the check measurements of the lines of the foundation should be taken before the concrete arrives. Timing of supervision is particularly critical where the work is subsequently to be covered up or enclosed.

3.2 Minimisation of waste

Wastage of materials occurs on all building sites. In some cases it cannot be avoided, and in others the measures to prevent it cost more than the materials saved, but in most instances careful attention to handling and storage can reduce the loss. The economics of waste must be carefully considered: clearly, if the builder only has need of, say, two cubic metres of aggregate on the job the cost of constructing a concrete storage bin would be more than the value of the non-useable aggregate at the bottom of a heap deposited on the open site. Similarly, the cost of a common brick dropped from a scaffold can be less than the cost of the bricklayer's time taken in retrieving it. However, the careful stacking of the bricks on the scaffold in the first place and their equally careful use should avoid dropping any at all and would, thereby, reduce both the wastage and the danger to men working below.

Dealing with materials has three phases, in each of which there is a potential for wastage:
1. *Delivery* – unless properly packed, loaded and held materials can be damaged in transit.
2. *Storage and protection* – incorrect storage and inadequate protection can result in materials which were correct when delivered being unusable when required or, worse still, failing to perform satisfactorily after installation.
3. *Issue and transportation* – the over-issue of materials gives rise to inevitable waste when the surplus is discarded and casual horizontal or vertical transportation is a very obvious source of material damage and loss.

3.3 Testing and storing concreting materials

3.3.1 Sand and aggregate

Both the sand and the aggregate used for concrete and mortar, being natural materials, are subject to variations and to the presence of harmful impurities. Materials obtained from a reputable pit are not likely to be sub-standard but quick and easy on-site tests and proper storage can help to avoid any possible trouble in the future. A sample of sand rubbed between the palms of the hand will readily reveal the presence of clay or silt which will be left on the skin and the process of rubbing will detect the sharpness of the grains. If either or both are unsatisfactory further tests are indicated.

If the sand is dirty, a fairly accurate measure of the amount of impurities can be achieved on site by the following method: Dissolve a teaspoonful of common salt in a pint of water and pour 50 ml of this solution into a 200 ml measuring cylinder. Add sufficient sand to reach the 100 ml mark and then pour in more water to reach the 150 ml mark. Shake the cylinder vigorously and then allow it to stand undisturbed for three hours. At the end of this time the separated silt will be visible as a layer on the top of the sand and the depth can be measured; it should not exceed 6 per cent of the depth of the sand. Any further tests to determine the percentage or nature of the impurities or the character of the grains are better undertaken in a laboratory.

Not only must the sand be clean but it must also be within the specified limits of grading of size of grains. If the grains are too large they will not interlock with the aggregate properly; if they are so small as to be of the nature of dust they will tend to replace the cement. This can be checked on site by spreading a sample very thinly on a board and examining it visually and if considered not in accordance with the specification, this can be followed by precise laboratory tests. The aggregate also must be clean and properly graded, both of which can be tested in the same way as sand. Whilst it is spread out on the board, the size of the largest and smallest particles can be checked against the specified size.

It is advisable to retain part of the tested aggregate and sand for comparison with later deliveries not only with respect to maintenance of standards but also for consistency of colour especially for materials for mortar or fair faced concrete.

Storage of both must be related to the mixer position and the facilities for delivery. It must also be arranged so as not to interfere with the transportation of the wet concrete.

The required concrete output and the frequency of deliveries will determine the size of the storage bins, and their shape will be arranged to suit the method of loading the mixer.

3.3.2   Cement

Cement will deteriorate in damp conditions or if left for some time. It is essential that it is stored in such a way as to ensure that the condition of the cement is as it was when it left the cement works.

The material arrives on site in two forms, either in bulk in a container lorry or in bags. In the first case, the construction programme must require a large quantity of concrete, sufficient to merit a substantial on-site mixing plant incorporating a cement silo. This is a large storage drum into which the cement is delivered at the top and from which the cement is transferred to the mixer at the bottom. The silo is weather tight, thus avoiding the danger of damp conditions, and because it is filled from the top and emptied from the bottom the cement is used in the same order as delivered so that none is left to go stale.

The same principles must be observed with bags. They must be stored off the ground, preferably on a boarded platform, and either well covered with waterproof sheeting or, especially if the storage is to be for more than a few days, contained within a weatherproof shed. Storage of bags must be so arranged that they are used in the order of delivery. This is not always easy to achieve since the first ones are at the bottom of the heap or at the back of the shed. To resolve this, separate stacks should be arranged for each delivery and used in correct rotation or if it is intended to use one storage building there should be a door opposite to the end at which the deliveries are made, from which the cement can be collected.

There are several different types of cement and whilst it is possible to test that the type delivered is as specified and described, the facilities of a laboratory are required and since the production process is rigidly controlled the contents of each batch from a reputable manufacturer can be relied on to be as described. It is, however, prudent to examine this description to check that the correct type has been delivered as errors can occur in this respect and the various cement powders do not look any different.

3.3.3 Water

This main ingredient of concrete and mortar must be as carefully controlled as the others. Any impurities which find their way into the final mix will have an inevitable adverse effect either on the strength of the concrete or on its appearance or on both, and these impurities can be contained in the water as readily as in the sand or aggregate.

The source of the water must be pure: this is not usually a problem in areas where a piped supply is available but must be carefully investigated in remote sites drawing water from other sources.

With a mixing plant the water is usually contained within a vessel and piped directly into the mixer. There is, in consequence, little chance of impurities getting in but on many sites it is stored in a large drum or tank which presents a very useful bath for the washing of tools, etc. This must not be allowed to happen. A spade recently used for digging or a trowel which the plasterer has just finished with will, if washed in the water butt, leave behind clay or silt or calcium salts all of which are harmful to the finished concrete.

3.3.4 Mixing

The testing and storage of the constituent materials of concrete is essential to the quality control of the finished concrete but equally so is the way these materials are combined and handled.

Maintenance of the correct proportions, by whatever method is used for measuring them, is very important, to ensure not only that the design strength is achieved but also that the strength is consistent throughout the structure. Perfect concrete results from the quantity of fine aggregate being such as just fills the spaces or interstices between the particles of the coarse aggregate and the quantity of cement just fills the interstices in the sand. This very dense, very strong material can only be produced under laboratory conditions where the volume of the interstices can be measured for each batch of materials, but the aim is to approach this standard under site conditions by careful specification of the grading and proportion of the materials.

The quantity of water also affects the finished quality and strength of the concrete. Generally, the more water the less the strength but the wetness of the mix is also dictated by the method of handling and the type of use to which it is to be put. Concrete which is to be pumped must, obviously, be more sloppy than a mix intended to be moved by dumper.

Testing the quantity of water (water/cement ratio) obviously must be carried out on site when the concrete is freshly mixed, and the method by which this can be done is the relatively simple slump test as shown in Fig. 3.1

3.3.5 Transportation and placing

If wet concrete is improperly handled the constituents of the mix can segregate, the heavier aggregate settling to the bottom and the lighter sand and cement rising to the top. This, clearly, would result in concrete of a very uneven quality if allowed to happen.

There are two ways in which this can occur: excessive shaking or vibration during transport (the drum of a ready-mix lorry rotates slowly but continuously during its journey to prevent this) and by dropping the concrete into position. The constituents separate when the concrete is dropped for the simple reason that the heavier particles continue to travel downwards after the lighter ones have stopped thus producing the same situation as excessive vibration.

Although excessive vibration during transport will have a harmful effect on the concrete, the right amount of vibration of the mix in position will produce the beneficial result of releasing air trapped in the deposited concrete. Releasing this air compacts the concrete, thus increasing its density, resistance to chemical attack, penetration of water vapour and frost and its strength. With regard to this last benefit, 1 per cent of entrapped air causes a reduction of strength of about 5 per cent.

Vibration is best carried out by means of mechanical vibrators such as vibrating pokers immersed in the mix, external vibrators clamped to or held against the outside of the formwork, beam vibrators used for compacting horizontal slabs and table vibrators onto which the mould boxes of pre-cast units are clamped. With small quantities of concrete, the air may be released by hand-tamping rods worked up and down in the mix.

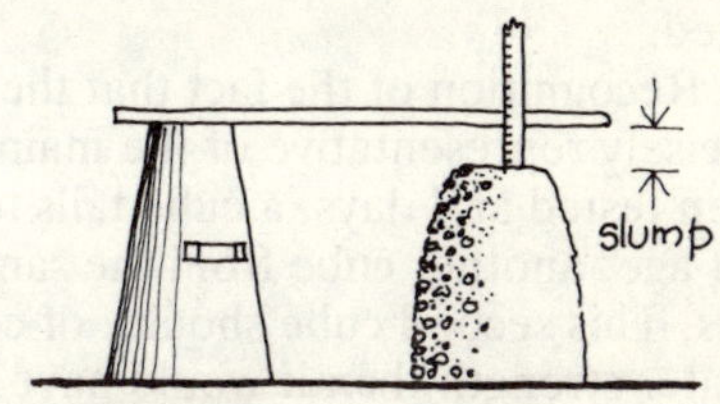

Type of concrete :	Slump m.m.
Vibrated very high strength	0
Vibrated high strength and mass concrete	0 - 25
Hand compacted mass concrete	25 - 50
Vibrated normally reinforced	25 - 50
Vibrated heavily reinforced	50 - 100
Hand compacted normally reinforced	50 - 100
Hand compacted heavily reinforced	100 - 150

Fig. 3.1 Slump test

3.3.6 Testing concrete

BS 1881 1970 describes tests and the apparatus required. The first of
these is the slump test, already described in Fig. 3.1, to determine the
water/cement ratio and workability of the wet mix. A series of slump
tests of each batch will demonstrate the degree of consistency of the
proportion of water but if the series also encompasses a fresh delivery
of aggregate, a change in the amount of slump may be caused by small

differences in the shape of the aggregate or its grading.

To test the concrete in its finished state it is necessary to take samples from the concrete to be placed in each part of the work, form each sample into a cube and test the compressive strength of the cube in a crushing machine. Compression tests can be either carried out before the work is begun, on a representative mixture of the sand, aggregate and cement which the builder is proposing to use; or on a works cube which is a sample of concrete taken from the fresh mix.

In the case of the works cube care must be taken to ensure that this small quantity of concrete is not allowed to be dried rapidly thus causing its performance under test to vary from the main mass of concrete which will cure more slowly. For this reason, the moulds containing the cubes must be kept under damp sacks for 24 hours before the specimens are removed; they must then be covered with damp material whilst being transported to the testing laboratory and on arrival must be stored in water again for 24 hours before being tested.

Recognition of the fact that the small test cube may not be precisely representative of the main mass is shown in the ruling that if, when tested at 7 days, a cube fails to attain the strength specified for that age, another cube from the same concrete may be tested at 28 days. This second cube should, of course, be expected to attain a greater strength than it would have done at 7 days. If, however, it still falls short of the specified standard the contractor may be given the opportunity of testing core samples taken from the actual concrete work as carried out to see whether the required strength has been achieved in the main mass.

3.3.7 Reinforcement

Site or laboratory testing of reinforcement materials is not necessary provided that they have been produced to comply with the relevant standards:

Steel wire for prestressed concrete	BS 5896 : 1980
Hot rolled steel bars	BS 4449 : 1978
Cold worked steel bars	BS 4461 : 1978
Hard drawn mild steel wire	BS 4482 : 1969
Steel fabric	BS 4483 : 1969

Storage and workmanship are, however, subjects requiring careful control. All reinforcement steel and made-up cages must be stored in dry conditions and carefully examined for loose rust, scale, oil or grease, any of which can prevent the adhesion of the steel to the concrete and cause failure of the structural member.

Reinforcement cages must be accurately assembled and adequately fastened together with galvanised wire ties, the cut ends of which must be turned inwards since if they are left protruding they will be near enough to the surface of the concrete to produce brown rust stains in exposed work.

Accurate positioning of all forms of reinforcement, whether individual bars, welded steel fabric or made-up reinforcing cages, is essential to ensure that the stresses are correctly distributed and that the steel is provided with an adequate cover of concrete. This can be achieved by the use of spacers either of concrete or plastic slotted onto or clipped to the bars. The practice of supporting reinforcement on odd bits of brick and short ends of timber should not be allowed.

A final check to ensure that nothing has moved and no unwanted objects have found their way into the formwork should be made immediately before the reinforcement is irrevocably encased in concrete.

3.4 Inspecting and storing bricks

Bricks are a bulky material, heavy, of relatively low cost and manufactured to a widely varying standard. However, this does not mean that the way they are handled and stored should be any the less careful than for other materials. The improper handling of facing bricks can damage the face or the arrises thus impairing the appearance of the finished wall. Carelessness with common bricks can, at worst, give rise to invisible partial fractures which can affect the strength of the finished work or, at best, result in unnecessary wastage; and incorrect stacking of either creates an untidy site and even dangerous conditions for anyone in the area. Figure 3.2 shows bricks stacked so that there is no danger either of the stack falling over or a single brick dropping onto a passer-by. Bricks often arrive on pallets or banded together with steel straps, in which case they can be stored as delivered until needed.

Bricks should be inspected for consistency of the load within the limits of variation for that particular brick and for consistency with

Fig. 3.2 Bricks stacked on site

previous deliveries. They should also be checked for soundness, which can be done by striking a sample brick with a trowel – if it is well burnt and sound it will give off a clear metallic ring.

BS 3921 : 1974 *Clay Bricks and Blocks* gives a full description and classification of bricks, BS 187 : 1978 describes six classes of calcium silicate bricks, more popularly referred to as sandlime bricks and BS 6073 Pt1 : 1981 defines solid, perforated, hollow and cellular concrete bricks. All Standards set acceptable levels of strength, water absorption and resistance to attack by chemicals and frost.

In the selection of samples for testing care must be taken, especially with clay bricks, to make the range as representative as possible. Clay bricks can vary widely in their characteristics and tests on isolated samples can be misleading. Sandlime and concrete bricks are more readily controlled with accuracy during manufacture and, as a result, are much more consistent in their properties.

3.5   Inspecting and storing blocks

A clay building unit more than 337.5 mm long, 225.0 mm wide or 112.5 mm high is, according to BS 3921 : 1974, a block, and can either be intended to be laid in mortar as a wall or used as a filler block in a reinforced concrete floor. The most common blocks on site now are light-weight or dense concrete but hollow clay blocks as described in BS 3921 are available and should be handled with as much, if not more, care than bricks to prevent damage and waste. They should be examined for uniformity of colour and squareness and absence of twisting or bowing and should be stacked on edge.

Most are grooved or scratched to provide a key for plastering but some are high-precision fair-faced blocks intended to be laid with care and simply painted. These require particularly careful handling to avoid damage to the arrises. On delivery blocks should be checked against the specification as to type and size. They should be stored on edge on a dry surface or on boards or old blocks to keep them off the ground and if they are in the open should be covered with polythene or a tarpaulin to keep out the rain.

3.6   Inspecting and storing timber

Timber, being a natural material, is subject to wide variation in quality and strength. Differences in appearance are one of the attractions of the material when used decoratively but differences in structural performance are not desirable and must be limited as far as possible. Careful visual inspection of the timber can eliminate the pieces likely to be weak and thus allow the rest to be stressed more highly. In the past the problem of variation in strength was solved by adopting a very low value for the stress carrying capacity. The modern practice is to grade the timber more precisely and thereby allow it to be used as a material with predictable properties. BS 4978 : 1973 and CP 112 : 1971 together provide the rules for stress grading either by inspection and measurement of the size and position of defects (visual stress grading) or alternatively machine stress grading is employed whereby each length of timber is passed through a machine which measures its resistance to bending and hence assesses its strength.

Timber is graded visually as GS (General Structural) or SS (Special Structural) and must be marked on at least one face edge or end with the graders' identifying mark and the grade of the piece. The machine grades are the same as visual grades but with M prefixed – MGS and MSS – and each piece is to be marked with the licence number of the grading machine, the grade of the piece, the British Standards Institution Kite Mark and the British Standard number BS 4978.

This is all done in the timber yard but the material should be checked again when it arrives on site to make sure no defective timbers have slipped through and that no damage has occurred in transit.

Common timber defects are:

Large or dead knots	— these cause weakness and must be avoided.
Shakes	— splits along the timber either radiating from the heart or following the growth rings around. CP 112 defines the permitted limits for shakes.
Wane	— a fault arising if, when the log was converted into boards, the surface of the log formed one or more of the arrises, i.e. one of the angles was rounded off. This reduces the cross-sectional area and hence the strength of the timber and is unsightly in timbers left exposed.
Blue stain	— this is frequently found in the sapwood of softwood, and is caused by a fungus. It does not affect the strength but the timber should be rejected if it is required for a natural finish.
Pin holes	— occasionally these are found in mahogany and similar timbers, and are caused by wood-boring insects attacking the timber whilst growing. The chance that it is an active infestation is remote but the defect looks unsightly in a polished surface.

Timber is frequently treated with a preservative against fungal attack or beetle infestation. If a colourless preservative is used, the timber should be accompanied by a certificate which should be examined. With carcassing timbers the preservative is coloured to show that it has been treated, and it is usually a fugitive dye which disappears after a short time.

All timber to be painted should be primed before delivery to site, and whether primed or not it should be stacked off the ground and sheeted over to keep it as dry as possible.

3.7 Inspecting and storing joinery

Many of the points in the preceding section also apply to timber frames, though additional careful inspection should be made to check on damage caused in transit which is much more noticeable in this work than in carcassing timber.

Frames should be stored on edge, preferably in a dry shed but certainly off the ground and under a tarpaulin. To reduce the risk of damage and moisture absorption it should be arranged that the joinery is not delivered to site long in advance of when it is required.

Care should be taken to see that all cut faces where window sills have been trimmed and horns removed are primed before the frames are built in.

3.8 Inspecting and storing boards

There are an increasing number of composite boards now being produced with enhanced properties, but the majority still to be found on sites are plywood, blockboard, chipboard and hardboard. Plywood is manufactured to BS 1455 : 1972 and the bonding between veneers is to BS 1203 : 1979; it is commonly available as three-ply 4, 6 or 9 mm thick or five-ply 9 or 12 mm thick. It is graded according to the quality of the face veneer and by the bonding material used.

There are three grades of face veneer: Grade 1 is the best, suitable for work to be polished, Grade 2 forms a base for good quality painted work and Grade 3 is for use where the surface is hidden. Of the four bonding agents used, only the first three, type WBP (weather and boil proof), type BR (boil resistant) and type MR (moisture resistant) are used for permanent work. The fourth, type INT (interior), is only resistant to cold water, not to micro-organisms, and should only be used for temporary work.

To check the quality, all boards manufactured to BS 1455 must be marked on the back, near one edge with the manufacturer's name or mark; the country of manufacture; BS 1455; the grade of face and back veneer (e.g. 2–3); bonding (WBP, BR, MR or INT) and nominal thickness.

Blockboard and laminboard, which are very similar, should conform to BS 3444 : 1972 and are graded, like plywood, according to the face veneers and bonding. The veneers are graded as Grade S (Specially Selected) or Grade 1 or 2, Grade 1 being suitable as an exposed face and 2 as a backing or base for painting. There are three bonding agents, corresponding to those used for plywood and these are type BR, type MR and type INT. As with plywood, the back or one edge must be marked with the information about the board set out above.

British Standard 5669 : 1979 specifies four types of wood chipboard: type I, Standard; type II, Flooring; type III, Improved moisture resistance and type II/III, Combined properties of II and III. All boards are to be marked 'BS 5669' followed by the maker's name or trade mark, the type of board and a coloured coding stripe, 25 mm wide on diagonally opposite edges – black for type I, red for type II, green for type III and red/green for type II/III.

Hardboard should be made in accordance with BS 1142 Pt 2 : 1971 in three grades: standard, super and tempered and is usually marked on the back.

All boards must be stored flat, supported over the whole of their underside clear of the floor in a dry warm store. If they are left on edge leaning against a wall they will take on a bend which is difficult to eradicate when fixing.

3.9   Inspecting and storing plastering materials

Plaster is made from lime, cement or gypsum, either singly or in careful combination, and with or without sand. All these must be checked when each is delivered to site to see that the first three are in accordance with the specification and that the sand is clean and dry.

The quality of the sand is very important since this is one of the finishing trades and any blemishes which can occur due to the presence of impurities cannot be covered up. Once delivered all materials, including the sand, must be stored in a clean, dry, covered store free from contact with the ground.

Delivery should be carefully timed to limit the period of on-site storage since the absorption of any moisture will shorten setting times and may reduce the finished strength of the plaster.

Water for plastering purposes must be kept absolutely clean; it must be separate from the water used for concreting purposes to avoid any cement finding its way into the gypsum plaster and must not be used for tool washing as the incorporation of small quantities of hardened plaster can affect the behaviour of the plastering mix.

Plasterboard is available in a range of lengths and widths, two thicknesses and with various features such as foil or polythene film linings. On delivery it should be inspected to see that these variables are all correct. Whilst plasterboard must be carried on edge to avoid breakage it must not be stored that way, otherwise it will take on a bow which is difficult to eliminate entirely in fixing. Boards must be stacked horizontal on a large, flat, clean and dry platform, under cover and to a height not exceeding 900 mm.

Other plastering materials – nails, scrim, expanded metal, coving, etc. – must also be checked against the specification and stored with care in a dry position.

3.10   Inspecting and storing metalwork

The main group of metal components found on a building site is structural steelwork, and delivery to site will involve heavy vehicles and lifting gear which must be arranged in advance.

All steel members of British manufacture are marked in the web during rolling with the name and trade mark of the manufacturer. This can be taken as an assurance that the steel and the member are in accordance with BS 4 but it should still be examined for straightness (occasionally steel beams and stanchions become bent in transit and require correction before fixing) and cut ends should be smooth enough for priming paint to be properly applied, or machined true if the member has to bear onto another.

Whether or not the steelwork is primed before delivery is dependent on whether the steel is to be encased or left exposed (in the latter case it is better to leave the priming to the painting contractor to ensure compatability with the painting system) and whether friction grip bolts are used, in which case priming is omitted to ensure close contact with the steel surface. Primed steelwork should be examined for damage to the paint film and any areas where the film is loose.

Storage of steel members is frequently in the open on site: they should be stacked on timber bearers and unprimed steelwork must be protected from the rain. The location of the storage area must be carefully chosen to avoid transporting these heavy members too far when required at their point of erection.

Steel is also present on site in the form of concrete reinforcement – which is dealt with under section 3.3.7. The other metals on site are mainly aluminium and copper alloys and tend to be small items which must be checked against specified references and stored in secure conditions.

3.11   Inspecting and storing sanitaryware

The full inspection of sanitaryware on delivery to site is difficult because each item is carefully packed and liberally protected by adhesive tape and yet the consequential costs of replacing, say, a faulty bath can exceed the cost of the bath itself by as much as three times its price.

Many of the glazed ware items are heavy and awkward to handle but are easily chipped if knocked or dropped and must, therefore, be unloaded and moved with great care. Since they are invariably quite expensive and readily removable, they should be stored in a place of security, but since the building is usually in the later stages of construction this does not generally present difficulties – a room with a lockable door can be set aside for the purpose.

3.12   Inspecting and storing plumbing goods

All materials used by the plumber and heating engineer for internal and external work are covered by British Standards and should be inspected to ensure that they are marked as being of the correct description.

They are also one of the ranges of goods which are attractive to thieves, being small, light and valuable, and in consequence must be stored in conditions of strict security, accessible only to the foreman and the plumber.

3.13   Inspecting and storing electrical goods and ironmongery

In common with plumbing items, the electricians' fittings and cable and the joiners' locks, bolts and handles are the class of items which are subject to pilfering and must also be securely and separately stored.

Inspection of the goods on delivery would be to see that all items comply with relevant Standards, are in accordance with the specification or schedule and are compatible with the rest of the items.

3.14   Quality control of workmanship

Since the person doing the work is, or should be, an expert in his craft, the maintenance of good standards of workmanship is something which ought to happen naturally, but unfortunately does not always do so. Where good standards have to be set by a supervisor, who may be less skilled than the craftsman, it must be done with care and tact if it is to be successful. A craftsman, offended by an implied, or even stated, criticism of his abilities is not likely to be sympathetic to the designer's ideals and hoped-for standards.

Motivation of people is a subject of considerable importance and exhaustive research in management techniques should be studied by anybody in control of the quality of workmanship.

There are many ways of checking the manner in which the work has been carried out – the surreptitious pencil mark to check on the number of coats of paint, the observation of one upright against another beyond, the running of the hand over a finished surface, etc. – which are acquired by experience; but mainly the inspector must rely on his eye to detect failures or short-comings in the first place which can then be further checked by measurement or other appropriate tests.

Part II

Substructure

Chapter 4

Soil testing and underpinning

4.1   Soil investigation

The word 'soil' describes the surface of the earth and is divisible into
'top-soil' which supports plant life but will not support building loads
and 'sub-soil' which does not contain the nourishment to sustain plant
life and, largely because of the absence of these plant foods, possesses
the capacity to sustain building foundations. In practice the prefix 'top'
or 'sub' is frequently dropped as the type of soil is clear from the
context in which it is placed.

It is obvious to even a casual observer that the nature of top-soil
varies greatly. So equally does sub-soil, and to ensure the safety of the
building it is necessary to investigate and test the sub-soil to measure
all the characteristics which determine its ability to carry loads and to
remain stable. There are many soil tests which can be carried out and a
foundation designer will only be interested in those which produce
results related to the soil characteristics which interest him.

4.2   Soil characteristics

There are four main sets of characteristics:
  (i) Physical properties
 (ii) Particle size
(iii) Chemical composition
(iv) Geological origin

The first two are closely associated with each other and with the performance of the soil under stress; the third is of concern to the designer anxious to avoid unfortunate chemical reactions with building materials. The fourth set is of little relevance.

All soils are defined in BS 1377 : *Methods of Testing Soils for Civil Engineering Purposes*, by particle size as coarse-grained or fine-grained, the former being non-cohesive, highly permeable and only slightly compressible (which occurs almost as soon as the load is applied), whereas the latter is the opposite – cohesive, impermeable and compressible over a long period of time.

BS 1377 relates soil names and particle size as follows:

Clay	—	less than 0.002 mm
Silt	—	0.002 to 0.06 mm
Sand	—	0.06 to 2.0 mm
Gravel	—	2.0 to 60.0 mm
Cobbles	—	6.0 to 200.0 mm

fine grained (Clay, Silt)
coarse grained (Sand, Gravel, Cobbles)

Positive classification of clays by particle size alone is difficult and therefore the way these behave in relation to their moisture content is used. When a lot of moisture is present the soil is in a liquid state, the clay particles being suspended in water. With less water the soil is in a plastic state and a further decrease leads to the solid state. The points of change are referred to as the plastic limit and the liquid limit and can be found by applying tests set out in BS 1377 (see section 4.4).

4.3 On-site tests

The overriding purpose of soil investigation is to predict the way the soil will behave under load – not the soil generally but those particular little areas on which the load is to be imposed – and for this reason a number of on-site tests have been devised to examine the soil to be loaded, in its natural conditions. Many experienced builders, architects and engineers will assess the bearing capacity of the soil in the bottom of a trial hole or trench by visual examination and the impression their heels leave in the surface. This is all very well if the person is experienced and knows the district intimately and if the loads to be dealt with are small – the sort of circumstances which can occur with infill housing for instance. Where the loads to be handled are large and the soil less reliable more precise methods must be employed, but nonetheless the critical examination by an experienced eye should not be ignored.

To investigate the soil it is necessary to get to it and for this purpose trial pits are dug or holes are bored. Since the information needed is about the soil below the foundations, the pits or holes must be taken to a lower level than the underside of the proposed substructure. This weakens the soil and therefore these pits or holes should not be taken out precisely where the foundation will be but as near as is practicable or necessary to give a correct indication. Better visual examination is achieved with trial pits which are usually in the region of 600 mm wide by 1200 mm long and 900 to 1200 mm deep but bore holes can be taken down to 30 m with modern equipment to reveal the nature of the ground at the lower levels.

4.3.1 Visual examination

Observation of the soil exposed in the sides of a trial pit will reveal the type and distribution of material, whether it is natural or infilled ground and the pattern of the strata. This examination must be carried out as soon as possible after excavation because, in many instances, water will enter the hole. Information on ground water is vital to the foundation designer but its presence can obscure the sides of a trial pit.

Following a visual examination, further information can be gained by specific tests.

4.3.2 Plate load test

A direct way to find out how the soil will react to compressive load is to apply a pressure and that is the basis of this test which is fully described in BS 5930 : 1981 *British Standard Code of Practice for Site Investigations*. It consists of a steel plate between 300 and 1000 mm diam. being placed on the surface of the bottom of a trial pit and loaded with sandbags, bags of cement, concrete blocks, a tank of water or even a heavy vehicle. The load – called kentledge – is not usually

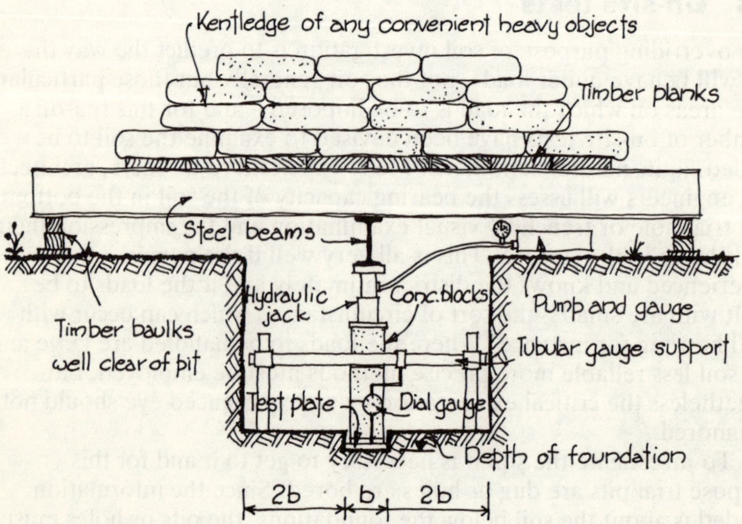

Fig. 4.1 Plate load test equipment

applied directly to the plate because of the problem it would present in accurate measurement and the sheer physical problem of getting, say, a sufficiently large pile of bags of cement onto a plate 600 mm diam. The kentledge is placed on a platform, supported well clear of the pit, against which an hydraulic jack acts to impose an accurately measured and progressively increased pressure on the plate. The pressure and the settlement of the plate are recorded and the test proceeds until the soil shears or a load developed equivalent to three times the building load (see Fig. 4.1).

By its directness this would seem to be a test method surpassing all others. Unfortunately, not only is it an expensive exercise but it can also be misleading because the test plate is generally smaller than the foundation and the bulb of distribution of the test pressure does not extend so deeply, as shown in Fig. 4.2, and therefore the results may be unaffected by underlying weaker strata.

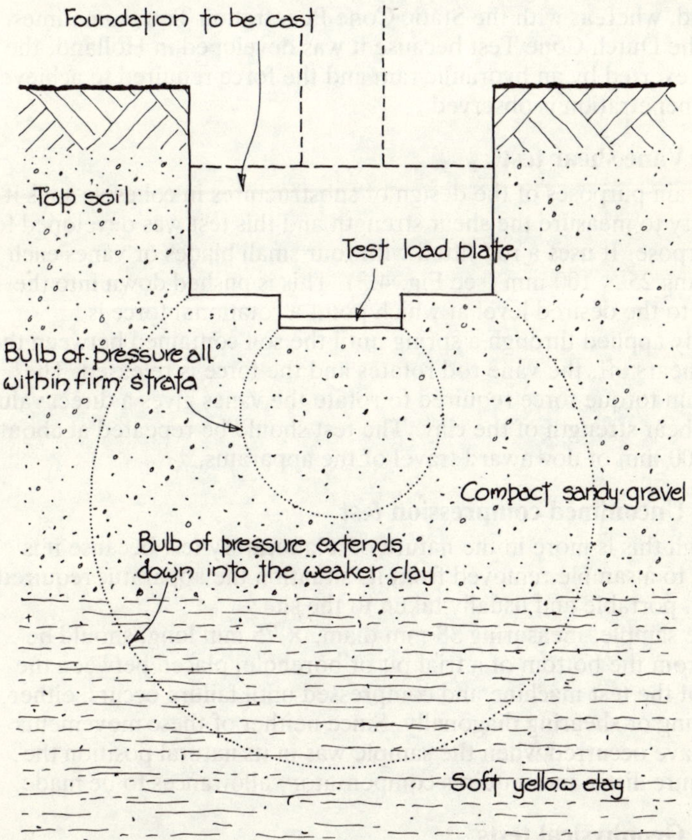

Fig. 4.2 Bulbs of pressure

4.3.3   Penetration tests

Another way to find out about the soil at low level is to drive
something down into the bottom of the pit or borehole and measure
how difficult this operation is. There are three forms of this type of
penetration test and generally they are more useful to the designer of
piled foundations than of spread foundations.

The Standard Penetration Test makes use of a sample tube fitted
with a driving shoe comprising a ring with a sharpened bottom edge.
The sampler is fitted onto a rod or rods and driven 450 mm down into
the ground. The number of blows required to achieve the last 300 mm
penetration is noted (the N-value) and from this figure the soil
resistance can be assessed. The sampler is then withdrawn and the
tube, which is split, is opened to allow examination of the contents.

The other two penetration tests employ a cone instead of a tube
and in the case of the Continuous Dynamic Cone Penetration Test it is
driven down into the ground by a drop hammer and the N-value
recorded, whereas with the Static Cone Penetration Test, sometimes
called the Dutch Cone Test because it was developed in Holland, the
force is exerted by an hydraulic ram and the force required to achieve
75 mm penetration is observed.

4.3.4   Vane shear test

For certain purposes of the design of substructures in cohesive soils it is
necessary to measure the shear strength and this test was developed for
that purpose. It uses a rod fitted with four small blades or vanes each
measuring 25×100 mm (see Fig. 4.3). This is pushed down into the
ground to the desired level at which point a rotational force is
gradually applied through a spring until the soil contained between the
vanes shears off, the vane rod rotates and the force is released. The
maximum torque force required to rotate the vanes gives a direct value
of the shear strength of the clay. The test should be repeated at about
every 300 mm of downward travel of the apparatus.

4.3.5   Unconfined compression test

Although this is more in the nature of a laboratory test because it is
applied to a sample removed from its situation the apparatus required
is small, portable and usually taken to the site.

The sample, measuring 38 mm diam. $\times$ 75 mm long, should be
taken from the bottom of a trial pit or borehole, placed between the
plates of the test machine and compressed until failure occurs, either
by bulging or shearing diagonally. Since neither of these movements
could have occurred when the sample was in its natural position the
test is unrealistic and requires compensatory allowances to be made.

4.3.6   Geophysical tests

Changes in the physical characteristics of soils and rocks can be
measured quickly and economically by geophysical methods but at the

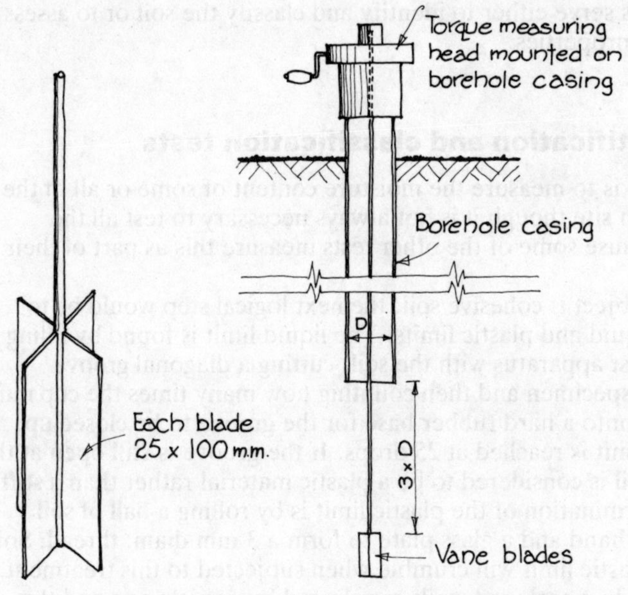

Fig. 4.3 Vane test equipment

present stage of development these are not a reliable substitute for the on-site investigative methods already described. They are more often employed where deep investigation of rock strata is required for, say, tunnel boring or where penetrative tests are difficult because of the presence of large stones in the ground.

The tests measure the electrical resistance between buried electrodes, the variations in the speed of travel of vibrations from a source through strata of differing density to a seismometer or the changes in the earth's magnetic field resulting from changes in the strata or due to features such as tunnels or mine shafts.

4.4 Laboratory tests

British Standard 1377 defines soil tests to determine the moisture content, the liquid and plastic limits, the particle size and distribution, the compressive and shear strengths, the permeability and the chemical constituents of a sample of soil.

The sample should be undisturbed for most tests, obtained by forcing a 100 mm diam. tube into the soil until it is full, removing it and sealing the contents by a cap at each end. To guard against possible changes occurring in the sample, tests should be carried out within two weeks of the sample being taken.

The tests serve either to identify and classify the soil or to assess its engineering properties.

4.5   Identification and classification tests

The first test is to measure the moisture content of some or all of the samples from site though it is not always necessary to test all the samples because some of the other tests measure this as part of their procedure.

If the subject is cohesive soil, the next logical step would be to assess the liquid and plastic limits. The liquid limit is found by filling a cup in the test apparatus with the soil, cutting a diagonal groove through the specimen and then counting how many times the cup must be dropped onto a hard rubber base for the groove to be closed up. The liquid limit is reached at 25 drops. If the groove is still open at this point, the soil is considered to be a plastic material rather than a stiff liquid. Determination of the plastic limit is by rolling a ball of soil between the hand and a glass plate to form a 3 mm diam. thread. Soils below the plastic limit will crumble when subjected to this treatment.

As already mentioned, soils are classed by particle size and this can be analysed, for the coarse grained soils, by weighing the amount of soil retained on each of a series of sieves of progressively finer mesh and from the results calculating the percentage of each. The fine grained soils cannot be tested in this way and for them the sedimentation method is used, whereby a specimen is stirred into distilled water to form a suspension. In accordance with Stokes Law, particles of different size settle at different velocities, and therefore a sample drawn off at a specified depth after a calculated time will only contain the particles too small to have descended further in the vessel. A later sample at the same depth will not contain the larger sized particles present in the earlier sample. The maximum particle size can be calculated from the depth and settlement time and the quantity deduced by weighing the residue in each sample taken. An alternative method to sampling the suspension is to take a series of measurements of the specific gravity of the suspension with a specially designed hydrometer since the specific gravity depends on the weight of the soil particles still in suspension at the time.

Once identified and classified the soil description should be stated in the sequence laid down in CP 2001, starting with consistency (for cohesive soils) or density (for non-cohesive soils), and then the structure, the colour and finally the particle size classification, giving the predominant soil type last, e.g.

consistency	*structure*	*colour*	*particle size class*
firm	laminated	yellow	sandy clay

4.6   Tests of engineering properties

After the laboratory worker has established the class to which the soil
samples belong he then tests them to measure their properties. There
are many such tests but the main ones are designed to establish the
permeability, the compressive strength and the shear strength of the
samples.

4.6.1   Permeability tests

The rate by which water flows through the soil gives an indication of its
load-bearing capacity and is also important where de-watering or soil
stabilisation systems are to be used or where land drainage, soak-aways
or sewage disposal systems are to be installed.

To obtain this information, the soil sample is placed in a cylinder –
possibly the sampling tube itself, and water run through it from a small
tank above. The volume of water flowing in a given time is measured
and from this can be calculated the coefficient of permeability which is
expressed as metres per second.

4.6.2   Compression tests

There are mainly two compression tests, the unconfined test, which
can also be carried out on site and is described in Section 4.3.5, or the
triaxial test, which, because of the size of the equipment, is strictly a
laboratory exercise.

As was mentioned in Section 4.3.5, failure of the sample under an
unconfined compression test is either by bulging or by diagonal
shearing, neither of which would happen in normal circumstances
because of the confinement of the surrounding soil. The triaxial test
eliminates this unrealistic situation by applying a horizontal restrictive
pressure all round the sample whilst a vertical load is applied. The
equipment in which this is done is shown in Fig. 4.4 and consists of a
rubber cylinder in which the soil sample is placed, mounted inside an
acrylic cylinder filled with water under a regulated pressure. A loading
ram applies a pressure to the top of the sample, the surrounding water
controls any tendency of the sample to bulge or shear diagonally and a
porous disc and pipe at the bottom of the sample allow the water
contained within the soil to be drawn off. The axial load is increased
slowly until a compression occurs. This test is repeated on a series of
samples with varying water pressures and the results plotted on a graph
from which the cohesion of the soil and the internal angle of friction
can be measured. These values give, respectively, the shear strength
and the angle of repose of the soil.

4.6.3   Shear strength test

The triaxial test gives a value for the shear strength of a soil but this
can be measured more directly by the equipment illustrated in Fig. 4.5

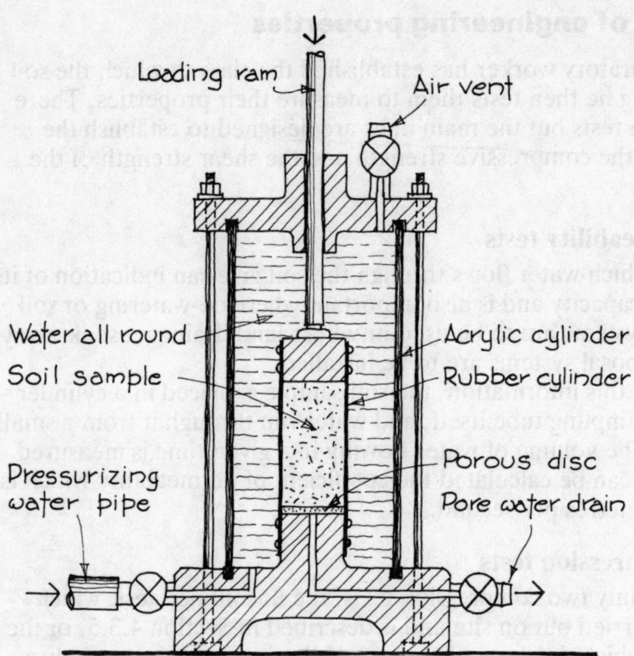

Fig. 4.4 Triaxial load test equipment

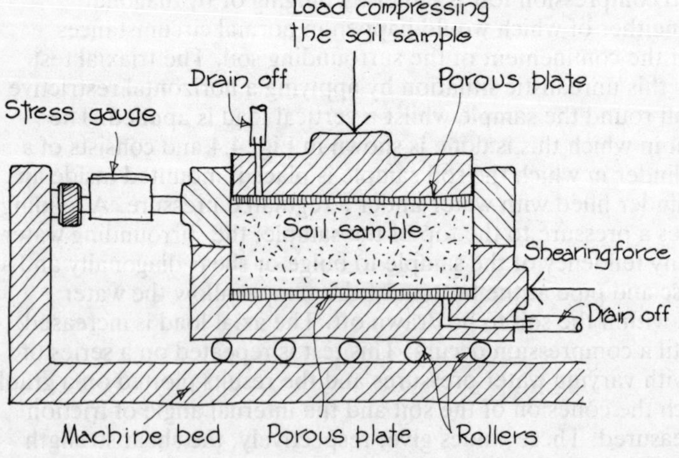

Fig. 4.5 Shear box equipment

in which a specimen of the soil is placed in a box which is split round the middle. Porous plates are placed at the top and the bottom of the box if the sample is wet, to allow drainage. A vertical load is applied to the top of the sample and a horizontal force applied to the upper and lower halves of the shear box. This force and the corresponding shear displacement are measured in a series of tests against different vertical loads and the results plotted on a graph from which the shear strength and angle of internal friction can be measured. The test is simple but, due to a number of disadvantages, is not so reliable or accurate as the triaxial test.

4.7 Underpinning

The primary objective of most soil investigations is to collect the information required for the design of foundations which will adequately sustain the building loads. Inadequate or incorrect information, errors in interpretation or application of the data can lead to foundation failure and the most expensive building maintenance operation – underpinning.

Foundation failure is not the sole reason for underpinning foundations. The necessity to do this can arise through work, either to the building itself or to an adjacent structure, which penetrates further into the ground than the bottom of the existing foundations and thus removes some of the soil which formerly supported that foundation (see Fig. 4.6).

There are several techniques employed, all with the same basic aim of taking the building loads deeper into the ground either to a stronger strata or to a level below an adjoining substructure. The most commonly employed is a continuous strip but piers or piles can be employed and, in favourable soils, cement injection can be economic alternatives.

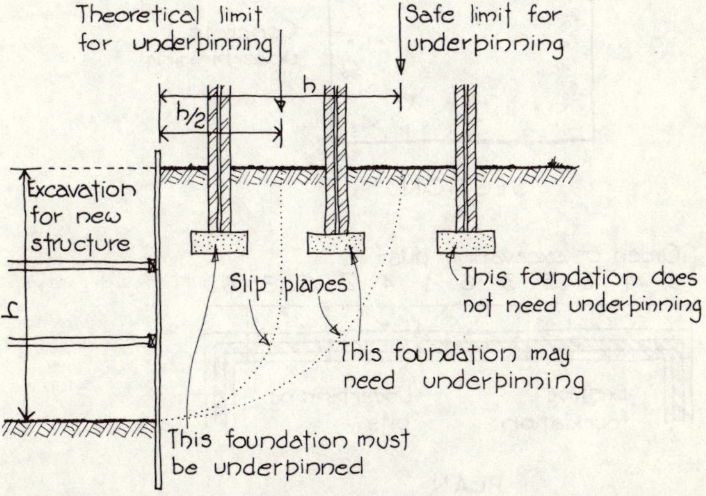

Fig. 4.6 New work causing underpinning

4.7.1   Continuous strip underpinning

This method can be used to correct faulty or inadequate foundations where soil of a greater load-bearing capacity is available reasonably near to the underside of the existing foundation or where the existing wall is to be carried further down into the ground to form a basement structure.

It consists of excavating in short lengths beside and below the foundation to the required level, filling the pit with concrete, firmed up to the underside of the foundation and then excavating and filling successive pits until a continuous concrete 'wall' has been formed. The order in which this should be done is shown in Fig. 4.7.

Great care must be taken in this operation because there is danger of building collapse and of trench collapse. If the soil is at all loose or

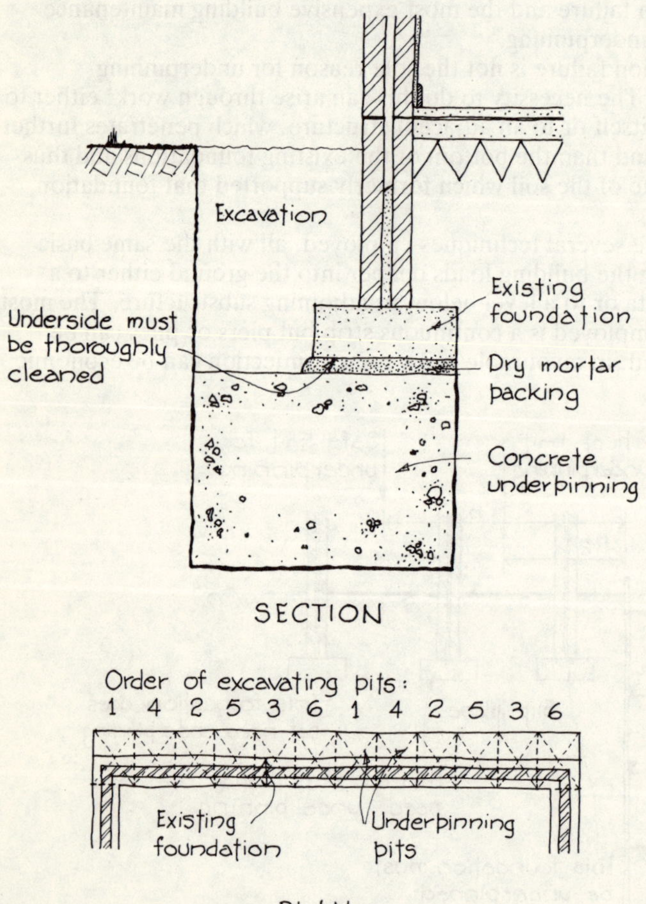

Fig. 4.7   Continuous strip underpinning

the excavation deep (both of which are quite likely) the pit sides must be supported by timbering or sheet steel piles. If the building shows signs of structural weakness the length of unsupported wall over each excavation must be kept shorter than the recommended maximum length of 1.2 to 1.5 m.

The pits should be set out evenly along the length of the wall and excavated in the order shown, pits with the same number being excavated at the same time and the bottom of each excavation being well consolidated prior to concreting. It is also important to check that all soil, particularly cohesive soil, is removed from the underside of the existing foundation: any shrinkable clay remaining between the old foundation and the new underpinning will dry out and allow the building to settle even further.

Concreting should take place as soon as possible to avoid deterioration or flooding of the pits and chases or pockets should be formed in the ends of each length to form a bond with the next section. These may occur naturally, formed by the walings used to support the trench timbering.

The concrete filling can be taken up to the level of the top of the old foundation to try to ensure that the wall is firmly supported but there is, however, a risk that the underpinning concrete will shrink, leaving a gap into which the old foundation will settle. A better method is to carry the underpinning up to within 50 to 100 mm from the underside of the old foundation, allow the new concrete to set and shrink and then pack the space with fine concrete or mortar mixed dry and well rammed in. This procedure slows down progress but assures a more satisfactory result.

4.7.2 Underpinning piers

The last method is only suitable for load-bearing walls. If the subject is a framed building, the use of underpinning piers is a more appropriate method as shown in Fig. 4.8. In this system the building stanchions are shored up (see next chapter) to relieve the foundation of any load, and the soil below the concrete base is excavated down to the required new level. To make this operation easier, and since it will become redundant anyway, the old concrete base may be broken up. Once the required level is reached a new concrete pad is cast and a concrete or brick pier is formed between it and the bottom end of the column. The detail of the top of the pier will depend on the existing structure but in the case of a steel stanchion provision must be made to refit the holding-down bolts if the old base has been broken up.

4.7.3 Underpinning piles

Piled underpinning can be used in the same circumstances as those outlined above, but generally as a group of four or six piles around the column with a reinforced concrete pile cap firmed up to the underside of the column.

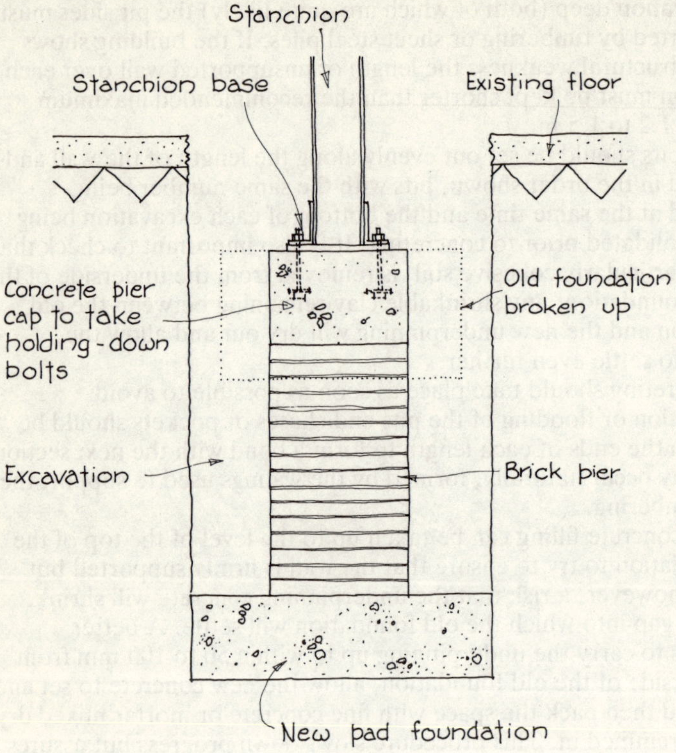

Fig. 4.8 Underpinning pier to stanchion

The use of piles is convenient if the safe bearing stratum is too deep for the excavation or if ground conditions are difficult, and can be employed to support walls as well as columns.

Bored rather than driven piles are generally used since they must be situated closer to the building than is practicable for a pile-driving rig, and furthermore the vibrations set up during driving could create serious problems in the existing structure.

Figure 4.9 shows various arrangements of piles for columns and walls. In the case of a system of columns to be underpinned, disturbance of the floor is inevitable since some of the columns will be within the building, but in the case of a building with external load-bearing walls only, preference is often given to an underpinning system which is totally from outside the building. For this reason the piles with cantilever caps find favour.

If the loads on the walls are relatively light and the bearing strata level such that the piles are fairly short, they can be placed sufficiently close together for the wall and its original foundation to span the distance without additional support. Conversely, with large loadings or

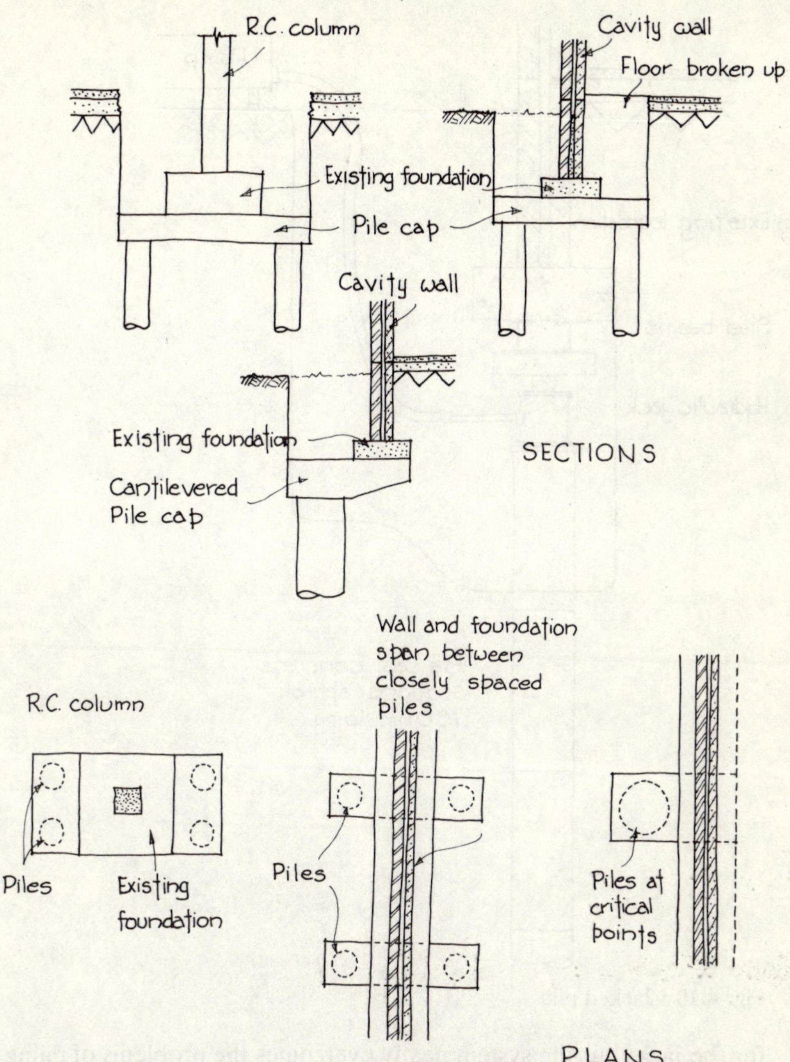

Fig. 4.9 Typical underpinning pile arrangements

widely spaced piles the wall has to be shored up and a reinforced concrete or concrete encased steel beam inserted to carry the loads to the piles.

There are a number of proprietary systems whereby a pit is excavated below the foundation and a pile in short sections bonded together is forced down into the ground by a jack or jacks acting against the underside of the foundation (see Fig. 4.10). Obviously, the mass of the building must be sufficient to provide an adequate reaction

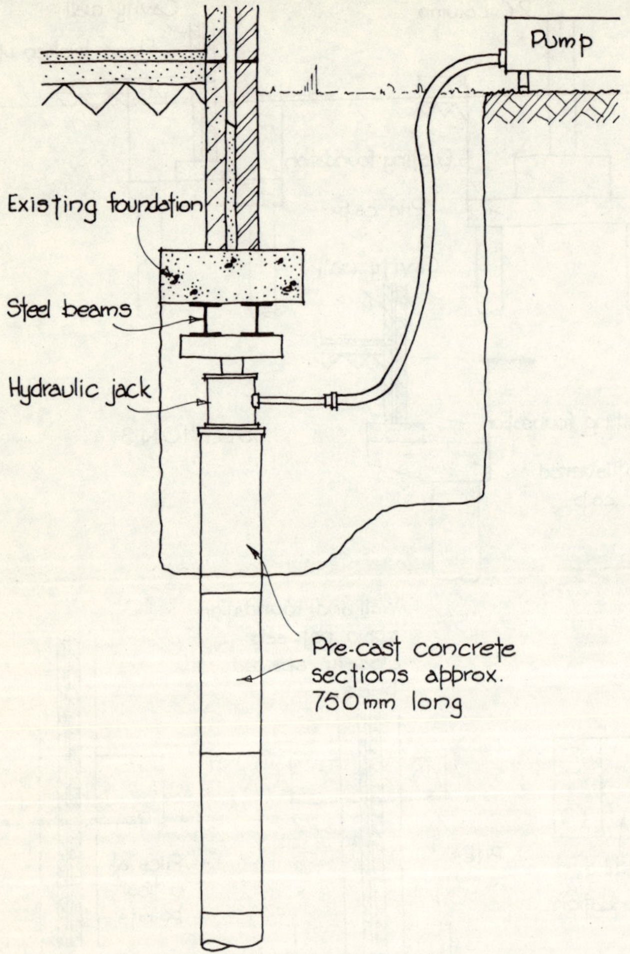

Fig. 4.10   Jacked pile

for the jacks but the system neatly overcomes the problems of piling below the foundation slab or strip.

4.7.4   Underpinning injections

Alternatively, cement or various chemicals can be injected into the ground beneath a foundation either to fill voids which have developed in the soil or to consolidate non-cohesive soils.

This method can be used to stabilise a building which has moved or to permit deep excavation to take place adjacent to the foundations. In the latter case open timbering with braces only would still be required, but not the close timbering or sheet steel piling which would otherwise be necessary.

Chapter 5

Temporary support

5.1 Work requiring temporary support

A building only requires the support described in this chapter if there exists a possibility that it will fall down. This possibility arises either because the support the structure has enjoyed has deteriorated to a dangerous extent or because the support is to be removed in the course of proposed building operations.

The loss of support due to deterioration arises, most commonly, as foundation failure due either to faulty assessment of the bearing capacity of the soil or to inaccurate or inadequate construction. The support will, firstly, be required to restrain the overturning of any walls which are moving out of plumb, and subsequently may be required in connection with remedial work to the foundation, such as underpinning.

Loss of foundation support may also be due to adjoining building works and can be dealt with by the provision of permanent underpinning; or alternatively temporary support may be provided when the loss is only for a short time.

By far the most common use of temporary support is as a provision to allow the removal of all or part of an element in the building fabric. Other works giving rise to the need for temporary support are demolitions where the removal of a building has weakened an adjoining structure, or where it is proposed to move a building bodily from one position to another when shoring will be required not only to support the building on the moving platform but also to prevent it distorting on its journey.

Whatever the reason, the provision of shoring is a highly skilled craft and should it be necessary it must be installed by someone with experience.

5.2 Points of support of restraint

The basic principle in the design of shoring is that the loads and stresses of a building can be considered, or found to be acting through specific points in the structure. The designer must identify these points and the angles of any inclined forces and devise a system of support which will contain them. Accurate analysis is essential, otherwise a shoring system may be erected for instance against the middle of a wall which is on the move only to find that the support just restrains the centre section whilst the two ends continue to collapse.

Clearly if the building is a framed structure, the points of support are the beams and columns. These are readily identified and will be all that require support, unless the trouble is that the infilling structure is leaving the frame, but this is usually fairly light and easily restrained. If, however, it is a brick or block structure, no convenient indications of concentration of load are given, nor indeed do they necessarily exist. Fortunately, the bonding of this type of work not only distributes imposed loads but also can be employed in reverse to concentrate distributed loads. If one brick is supported it will support two above it and they, in turn, will support three above them and so on, the amount of wall being supported increasing in each course until the bonding is interrupted by an opening or the end of the wall. Naturally this effect must be limited because the stress in the lowest, single, brick could exceed its load-carrying capacity, but it does allow supports to be inserted into walls at carefully selected intervals whilst, say, a large opening is formed.

It is also important to provide support to a wall where an upper floor load occurs, and for this reason the centre line of a raking support should pass through the point where the load is imposed (see Fig. 5.2).

5.3 Dead shores

These are the most common and are used whenever the permanent support is to be taken away. Each must be individually designed but all dead shores consist of a horizontal beam called a needle which takes the load and two struts, one each end, to transfer the load from the needle to the ground or other stable surface capable of carrying it.

Figure 5.1a shows a dead shoring system in a brick wall preparatory to cutting a large opening and it can be seen that it is necessary to ensure the struts have a firm foundation, even to the

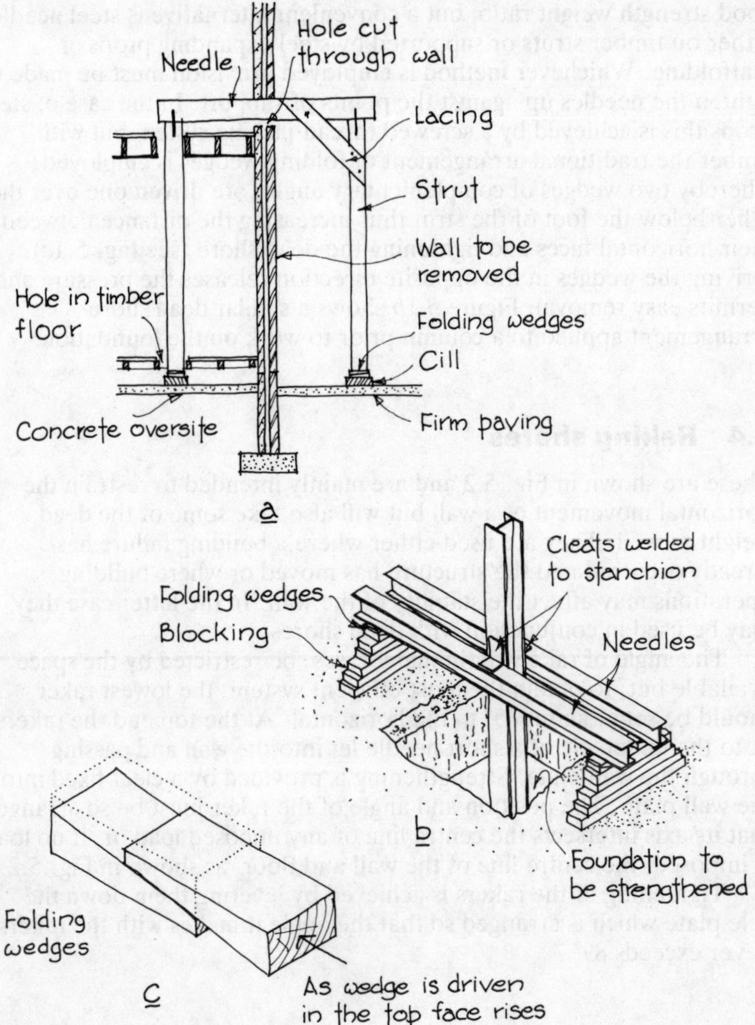

Fig. 5.1 Dead shoring

extent of removing some of the ground floor, if it is of suspended timber construction, to obtain a bearing on the oversite concrete below. To limit the length of the needle and thus reduce the bending stresses, the props should be as close as possible while allowing adequate working space. The spacing of the needles will be dictated by the loads to be supported and the nature and condition of the wall. The maximum distance apart is usually 2.0 m.

Figure 5.1 shows structural softwood needles and props and this is the material most commonly used because of the ease of fixing and a good strength/weight ratio, but a convenient alternative is steel needles either on timber struts or supported by steel expanding props or scaffolding. Whichever method is employed provision must be made to tighten the needles up against the points of support. In the case of steel props this is achieved by a screwed thread jacking system but with timber the traditional arrangement of folding wedges is employed, whereby two wedges of complementary angles are driven one over the other below the foot of the strut thus increasing the distance between their horizontal faces and tightening the dead shore (see Fig. 5.1*c*). Driving the wedges in the opposite direction releases the pressure and permits easy removal. Figure 5.1*b* shows a similar dead shore arrangement applied to a column prior to work on the foundation.

5.4 Raking shores

These are shown in Fig. 5.2 and are mainly intended to restrain the horizontal movement of a wall but will also take some of the dead weight as well. They are used either where a building failure has already occurred and the structure has moved or where building operations may affect the stability of the wall. In the latter case they may be used in conjunction with dead shores.

The angle of rake will, in many cases, be restricted by the space available but, to obtain the most efficient system, the lowest raker should be set at about 45° to the horizontal. At the top end the rakers fit to the underside of a short needle let into the wall and passing through the wall plate. Strengthening is provided by a cleat fixed into the wall plate. The position and angle of the raker must be so arranged that its axis intersects the centre line of any imposed load or, if no load is imposed, the centre line of the wall and floor, as shown in Fig. 5.2.

Tightening of the rakers is achieved by levering them down the sole plate which is arranged so that the angle it makes with the rakers never exceeds 89°.

5.5 Flying shores

When the need for temporary support arises because of the removal of a building in a row – such as the demolition of a mid-terrace house – or where the feet of raking shores would obstruct operations and a wall of adequate strength is convenient, flying shores should be used. This form of support is shown in Fig. 5.3 and is totally independent of the ground. For this reason this system can only restrain lateral movement and dead shores must be inserted if a vertical load is to be carried as well.

The length of the shore itself should be not more than about 10 m for practical reasons of installation and because of the limitation the

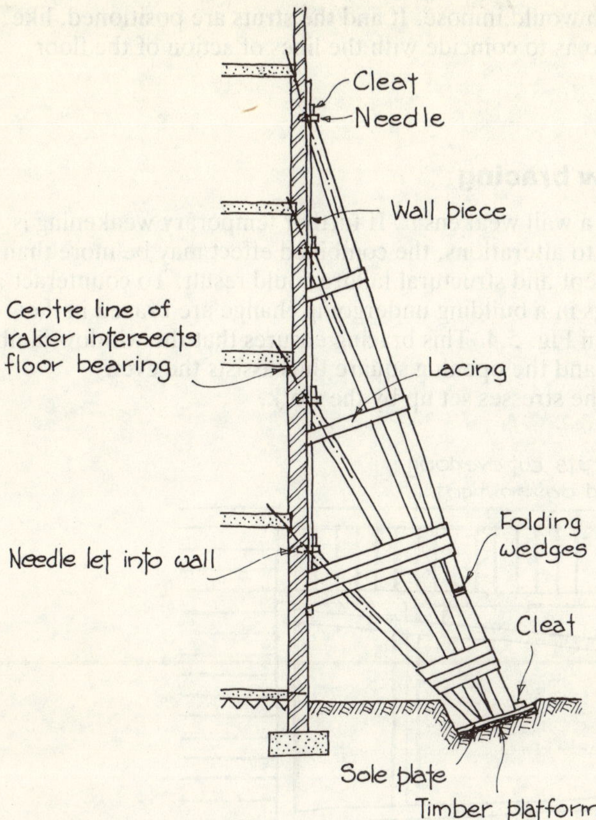

Fig. 5.2 Raking shore

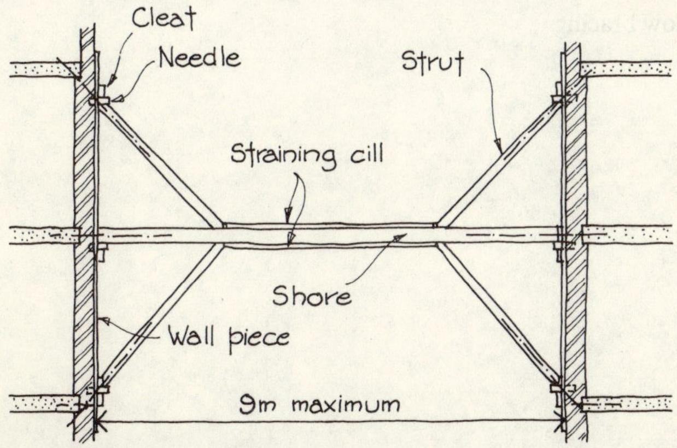

Fig. 5.3 Flying shore

slenderness ratio would impose. It and the struts are positioned, like raking shores, so as to coincide with the lines of action of the floor loads.

5.6 Window bracing

Any opening in a wall weakens it. If further temporary weakening is anticipated due to alterations, the combined effect may be more than the wall can accept and structural failure could result. To counteract this the openings in a building undergoing change are braced in the manner shown in Fig. 5.4. This bracing ensures that the window jambs remain parallel and the opening square thus assists the even distribution of the stresses set up by the work.

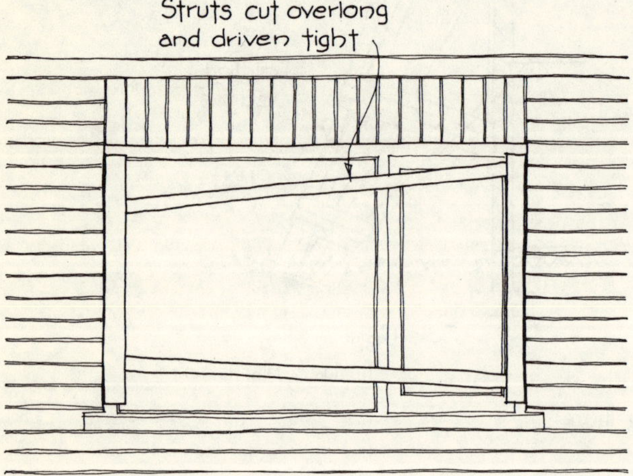

Fig. 5.4 Window bracing

Chapter 6

Choosing a foundation

6.1 General factors

There are, in principle, five types of foundations:

1. Strip foundations
2. Pad foundations
3. Raft foundations
4. Buoyancy or Box foundations
5. Piled foundations

Which particular type, and which method within the type, is finally adopted by the building design team will depend on four main factors.

6.1.1 Structural factors

The most obvious influence on the foundation designer when selecting the most appropriate system is the nature and structure of the building to be supported. Some are very light – air-supported structures need more holding down than support; some are flexible – many old timber structures demonstrate the extent of movement they can accept; some are rigid and will crack at the least sign of movement; others impose a continuous load, and yet others are designed with the load concentrated at specific points.

Assessment of these structural characteristics and the way they relate to the nature and bearing capacity of the ground are subjects for judgement by someone with technical knowledge or experience. Table 6.1 shows an extract from a table produced by the Cement and

Table 6.1 Choice of foundation.

Soil type and other site factors	Foundation type to use	Notes
1. Rock, hard sound chalk, sand and gravel	Strip foundation and trench fill	1. Minimum depth for protection against frost heave 450 mm for frost-susceptible soils. 2. Beware of swallow holes in chalk. 3. Keep base of trench above ground water level where possible.
2. Uniform firm and stiff clays: (a) No nearby trees	Strip foundation and trench fill	1. Depth to underside foundation must be as least 1 m.
(b) Where trees are near to building	Concrete piles and ground beams or Concrete piles and *in situ* R.C. floor or Specially designed trench fill	1. Clay type and tree distance will dictate piling detals 2. Trench fill can be used in low- to medium-shrinkage clay in the perimeter zone of tree root system. 3. Allow for void below floor for expansion of clay if floor is cast in very dry weather.
(c) Where trees are cut down just before foundation constructed	Concrete piles and ground beams	1. Piles must be long enough to resist clay heave. 2. Top of pile to be sleeved to reduce friction and uplift.

3. Soft clay Soft silty clay Soft sandy clay Soft silty sand	Wide strip footing or R.C. raft or Piles	1. Footings must be reinforced. 2. Settlement of footings and raft will occur. 3. Service entries must be flexible. 4. Piles taken down to firmer strata.
4. Peat	Concrete piles R.C. raft	1. Piles must penetrate to firm strata. 2. If peat layer is shallow dig it out and use wide strip footings or consolidated fill and raft. 3. If peat layer is thick (3–4 m) and firm with no firm strata below consider R.C. raft
5. Filled site	Concrete piles R.C. raft Wide strip Trench fill	1. Piles taken down to underlying firm strata. 2. R.C. raft and reinforced wide strip can be used if the fill was selected and well compacted. 3. Trench fill can be used if the depth of made ground is less than the depth of the trench. 4. Take precautions against combustion, toxic wastes and methane gas.

Concrete Association giving guidance on the choice of foundation for a house.

A site condition which can have a great influence on the choice of foundation type is the presence of any trees which are close enough to the building for the roots to affect the character of the ground (roughly a distance equal to the mature height of the tree). This condition can rule out the use of any shallow foundations which spread their load onto the ground above the level of the roots and it applies to sites where trees have been cut down shortly before building work started as well as to sites with surviving trees.

6.1.2 Design resources

The Building Regulations give 'deemed-to-satisfy' provisions for strip foundations[1] and thus if this system is adopted no design needs to be done and no calculations produced to the local authority. This can have the result that strip foundations are used, even in situations where another system may prove more economic or efficient but would have to be designed and proved by calculation. This is particularly applicable in the domestic sector of the industry. Elsewhere (and increasingly in housing work) experienced engineers are employed on foundation design and will produce a scheme using the most appropriate method.

6.1.3 Relative costs

Not only must the foundation design be shown to be the best technical solution, it must also offer the best economic solution when related to the contractor's expertise. For instance, short bored piles and ground beams can be shown to be competitive in price with strip foundations if the latter have to be taken down to 1.5 m or more but the actual economic result can also depend on the experience (and therefore the efficiency) of the firms involved in the use of the equipment and the organisation of the work on site, and also on the cost of obtaining and transporting the necessary piling rig.

Similarly trench fill foundations, although more concrete is used, can be less expensive than normal strip foundations because of savings in labour and the timber which would have been needed in shuttering if traditional trench support methods had been used. Modern hydraulically operated steel trench shuttering can alter this cost balance but only if the equipment is readily available and known to the men on site.

6.1.4 Working conditions

Any work below ground level must take into account the effect the condition of the ground can have on the method employed. A foundation system which can only be carried out with the assistance of heavy equipment is not a good choice on a very wet site of low bearing capacity where vehicles and transporters can get bogged down. A type of foundation system where this could apply is driven piling; and yet in some wet conditions, say during winter months, pile driving can continue at times when the construction of strip foundations, with its open trenches, would have to stop.

The effect of ground water in open excavations is a subject of considerable importance and where the ground water level is high the cost of de-watering can easily render uneconomic any system requiring such excavations.

Whilst ground water is a major controlling factor, others too may have a determining influence. For instance, the close proximity of existing buildings can preclude the use of driven piles because of the associated vibrations; or alternatively, the same foundation system may be ruled out because of difficulties in access to the site making it impossible or very awkward to manoeuvre the long pre-cast concrete piles required.

6.2   Settlement

The word 'settlement' is frequently used in relation to foundations whenever a building exhibits a crack, though often the rupture of the fabric has nothing to do with foundation movement at all and when it has it is not necessarily 'settlement'.

Almost every building constructed settles to some extent due to consolidation of the soil and the larger the building the more it will settle unless founded on rock. It is possible to estimate the amount from the results of a plate load test and it can be shown that, with a tall heavy building, downward movement of up to 150 mm must be allowed for. The skill of the foundation designer must be aimed at ensuring that the amount of movement is as nearly as possible the same at all points. The 'settlement' mentioned at the start of this section is actually 'differential settlement' resulting in one part of the structure moving more than another with inevitable results.

6.3   Types of continuous support

Building structures bring their loads down to the ground either along continuous lines or at isolated points. In the first case the substructure must achieve a load-carrying capability along continuous lines and in the second case it must achieve the same capability but at the points where the load is imposed.

Continuous support can be provided by:
1. Strip foundations
2. Raft foundations
3. Ground beams and piles
4. Buoyancy foundations

The constraints of the site and the circumstances arising from the building form and size will invariably dictate which type to select.

6.3.1 Strip foundations

The oldest and most common form of foundation is a strip foundation where a trench is excavated, concrete placed in the bottom and the wall built upon it. The depth is determined by the need to place the strip below the level where expansion due to frost will affect its stability (usually taken as 1 m) and the nature of the sub-soil. The width is governed by the relationship between the imposed load and the bearing capacity of the ground and also by the practical necessity of making it wide enough for a man to work in.

Regulation D7 of the Building Regulations contains a table giving minimum widths of strip foundations for various types of sub-soil and ranges of loading conditions between 20 and 70 kN/m of wall.[2] These minimum widths may need to be exceeded for the practical reasons mentioned above. In these circumstances the 'deep-strip' or 'trench fill' type of strip foundation can prove economic when it can be excavated by machine and filled with concrete thus avoiding the need for any workmen to descend into the trench at all. The quantity of concrete is greater but the volume of excavated material and amount of walling are both less. Figure 6.1 illustrates strip foundations and shows the graphical method for determining the depth of concrete required to resist shear.

6.3.2 Raft foundations

Sub-soil of low bearing capacity can require strip foundations to be of such a width that they almost meet. In these circumstances it is often decided to let them meet as a continuous slab under the building and reinforce it as a raft distributing the load on the soil over the whole area of the building. By this means differential settlement and structural failure is avoided.

Since the area over which the load is distributed is very large, the compressive stress on the soil can be reduced to the point where it can be carried by the strata quite near the surface; indeed, one of the advantages of a raft is the absence of deep excavations across the site. This shallowness, however, leaves the foundation vunerable to the effect of frost heave at the edges and hence it is often finished with an edge beam or toe extending down into the ground as a frost wall.

The system is frequently used for domestic work as shown in Fig. 6.2, but can be applied to larger structures, when the raft may be stiffened with beams to form a 'waffle' raft, also shown in Fig. 6.2.

6.3.3 Ground beams and piles

An alternative to a raft foundation where poor or unreliable ground conditions are encountered near the surface is to bore or drive piles down to firmer, more stable soil, excavate a shallow trench from pile to

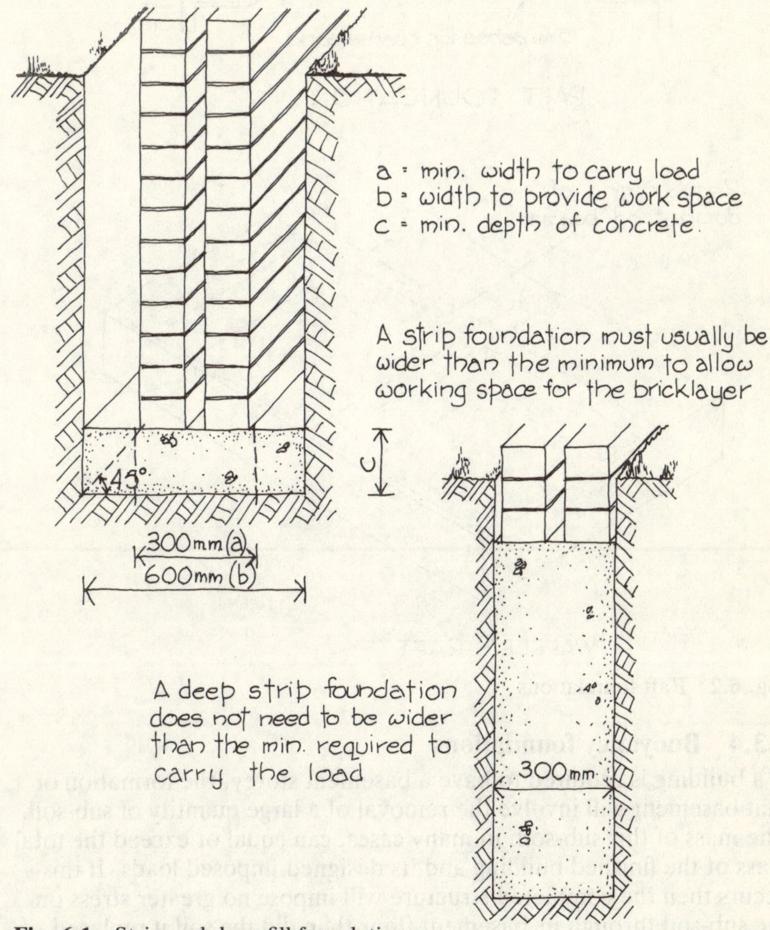

Fig. 6.1 Strip and deep fill foundations

pile and in it cast a reinforced concrete beam of sufficient strength to transfer the continuous load of the wall to the piles, as shown in Fig. 6.3.

This method is favoured by many foundation designers where the proximity of trees and the effect of their roots extracting moisture from the ground can cause failure of strip foundations. In domestic work this method, using hand-augered short bored piles, can be economically competitive with strip foundations when the latter have to be taken to depths greater than the minimum 1 m, and for other, larger projects it can offer the best solution of any.

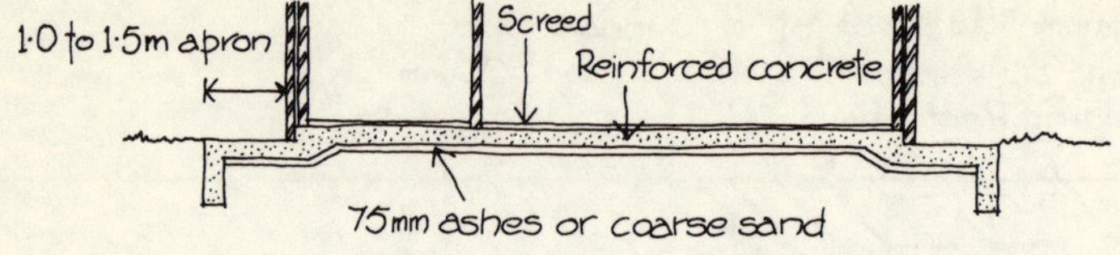

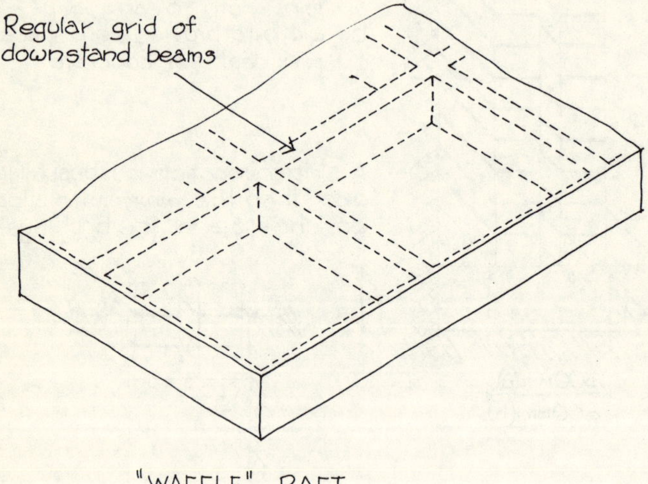

Fig. 6.2 Raft foundations

6.3.4 Buoyancy foundations

If a building is required to have a basement storey, the formation of
that basement will involve the removal of a large quantity of sub-soil.
The mass of this sub-soil, in many cases, can equal or exceed the total
mass of the finished building and its designed imposed loads. If this
occurs then the completed structure will impose no greater stress on
the sub-soil through its basement floor than did the soil it replaced.
Therefore the capacity of the soil to carry the load is not in doubt and
the building 'floats' in the ground precisely the same way that an ocean
liner floats in the sea (see Fig. 6.4).

As the sub-soil at the levels to which this form of substructure is
carried is almost always water-bearing the basement would need to be
lined with a waterproof membrane which would then cause the upward
hydrostatic pressure to counter-balance a proportion of the downward
thrust of the building.

The following simplified exercise demonstrates this principle:

Assume: sub-soil mass of 2400 kg/m^3
building mass of 1200 kg/m^2 of floor area
basement depth of 3 m

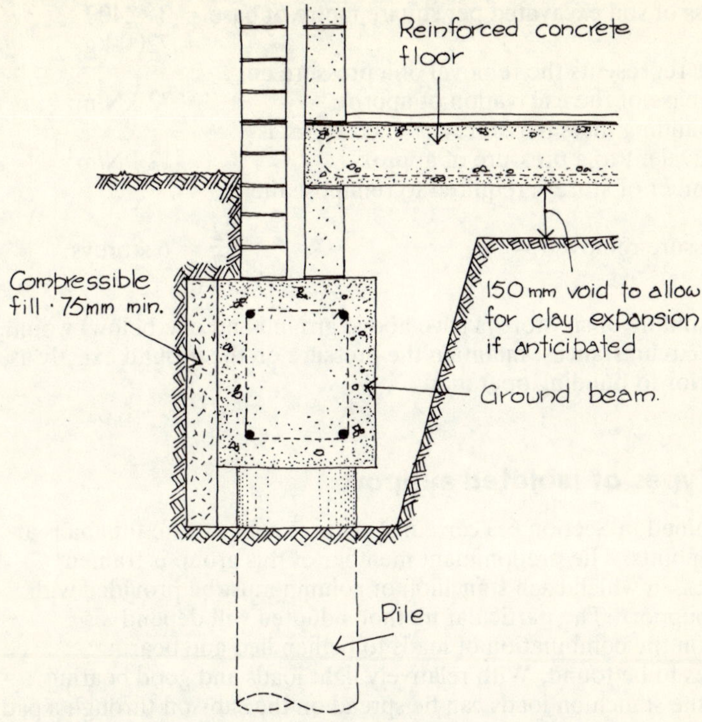

Fig. 6.3   Ground beam and pile foundation

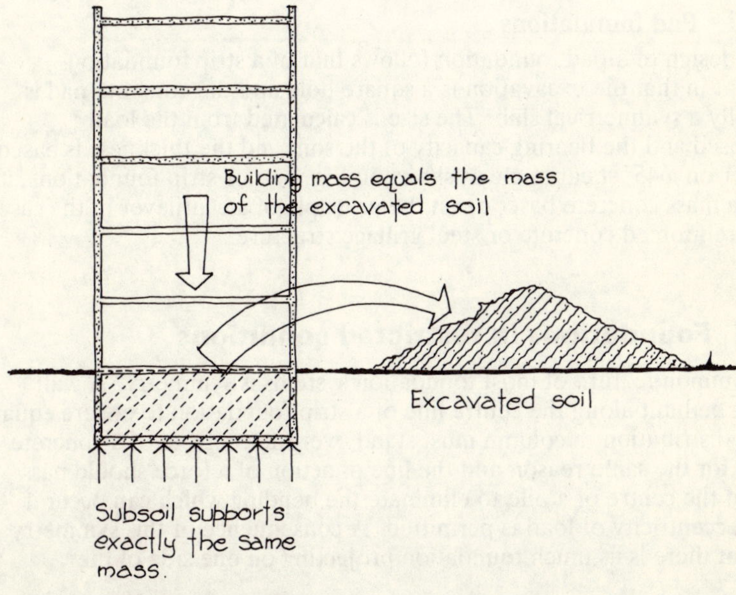

Fig. 6.4   Buoyancy foundation

Mass of soil excavated per square metre of base $= 3 \times 2400$
$$= 7200 \text{ kg.}$$

This represents the removal of a pressure on
the base of the excavation of approx. 72 kN/m^2
A building mass per storey of 1200 kg/m^2 is
equivalent to a pressure of approx. 12 kN/m^2
Number of storeys required to reimpose the

pressure removed $= \dfrac{72}{12} = 6$ storeys.

i.e. A building of six storeys (five above ground and one below) would,
in these circumstances, maintain the pressure on the ground exactly as
it was prior to building operations.

6.4 Types of isolated support

As explained in Section 6.3 certain structures concentrate the loads at
isolated points. The predominant member of this group is framed
structures, in which each stanchion or column must be provided with
its own support. The particular method adopted will depend, as
before, on the combination of loads to be handled and bearing
capacities to be found. With relatively light loads and good bearing
ground the stanchion loads can be spread on the sub-soil through a pad
foundation but in other cases piles, as described in Chapter 7, must be
employed to achieve the necessary strength.

6.4.1 Pad foundations

The design of a pad foundation follows that of a strip foundation
except in that the excavation is a square hole and the concrete pad is
usually a symmetrical slab. The size is calculated from the load
imposed and the bearing capacity of the soil, and the thickness is based
either on a 45° shear angle as shown in Fig. 6.1 for strip foundations, if
it is a mass concrete base, or on the principle of a cantilever in the case
of a reinforced concrete or steel grillage structure.

6.5 Foundations in restricted conditions

A common feature of most foundation systems is symmetry. A wall
must be built along the centre line of a strip foundation to ensure equal
load distribution, a column must stand over the centroid of a concrete
base for the same reason and the line of action of a force should pass
down the centre of a pile to eliminate the bending which can occur if
any eccentricity of load is permitted. A consequence of this symmetry
is that there is as much foundation projecting on one side of the

superstructure as on the other, which is all right if sufficient space exists to permit this but where the wall or column is at or close to the boundary of the site the space available can be insufficient or even non-existent.

One solution to this problem is to move the wall or column away from the boundary sufficiently far to provide room for the foundation. The wastefulness of this solution in terms of undeveloped site area and reduced building is obvious, and it also produces undesirable narrow strips of land which are usually inaccessible and impossible to maintain.

The alternative is to find ways of resolving the problem of support at these perimeter points or lines.

Placing a wall along the edge of a concrete strip produces a tendency for that strip to tilt as a result of the inequality of stress on the sub-soil. The simplest way to prevent this tilt is to bring the foundation structure up to the floor and tie the two together with reinforcement as shown in Fig. 6.5. By this means the floor acts as a membrane preventing the foundation from rotating and thus stabilising the support.

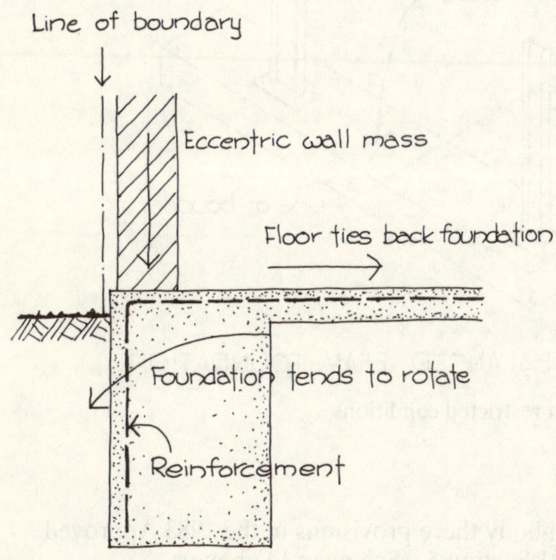

Fig. 6.5 Perimeter foundation

Columns at the boundary can be supported either on the end of a cantilevered foundation beam counterbalanced by the inner stanchions or else on the end of a balanced beam foundation as shown in Fig. 6.6 where two bases are combined to produce a beam in which the column loads are balanced each side of the centroid of the beam.

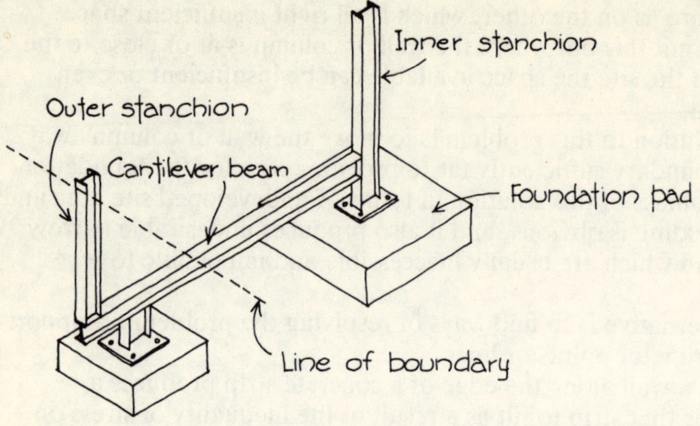

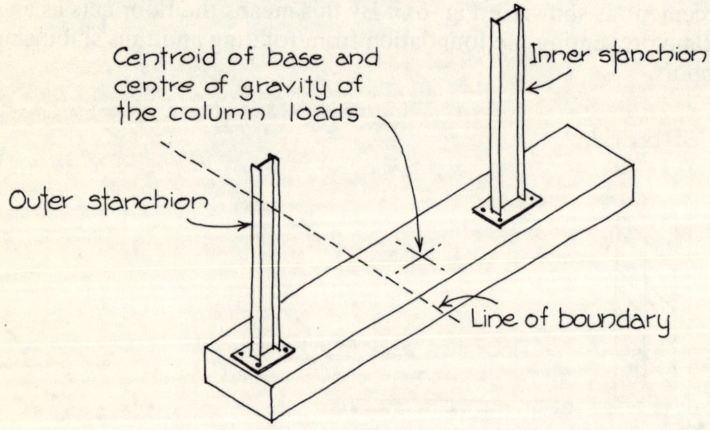

Fig. 6.6 Foundations in restricted conditions

Notes

1. It is proposed to embody these provisions in the 1983 Approved Document AD.B.01 Section 5. (See page 14 above)
2. See note 1

Chapter 7

Pile foundations

7.1 Principles of pile foundations

There are several types of pile – bearing piles which carry an axial
load, anchor piles which resist upward forces, sheet piles which form a
continuous membrane in the ground, piles to resist horizontal forces,
inclined piles, etc. In this chapter the first – bearing piles – will be
considered.

Driving stakes into the ground to obtain a stable base for a
subsequent structure is a practice of great antiquity and the only
differences between the prehistoric lake dweller's methods and
twentieth century practice is that we have substituted other materials
for the timber originally used and machines for muscle to drive them
in. Modern technology has also developed a technique of boring holes
and filling them with concrete as opposed to hammering a timber,
concrete or steel pile into the ground.

The decision to use piles as a foundation system will be influenced
mainly by the site investigation report and in most cases arises from the
lack of suitable bearing strata conveniently close to the underside of
the building. The type, condition and disposition of the lower layers of
sub-soil will dictate the way in which the pile obtains its load-carrying
capacity. Figure 7.1 shows two of the many sets of conditions which
can be found. In Fig. 7.1*a* there is the situation where a soft sub-soil
overlays a stronger, firmer lower strata in which case the pile gains
most of its support by end bearing; whereas the pile shown in Fig. 7.1*b*
passes through stiff clay and is mainly held up by the frictional

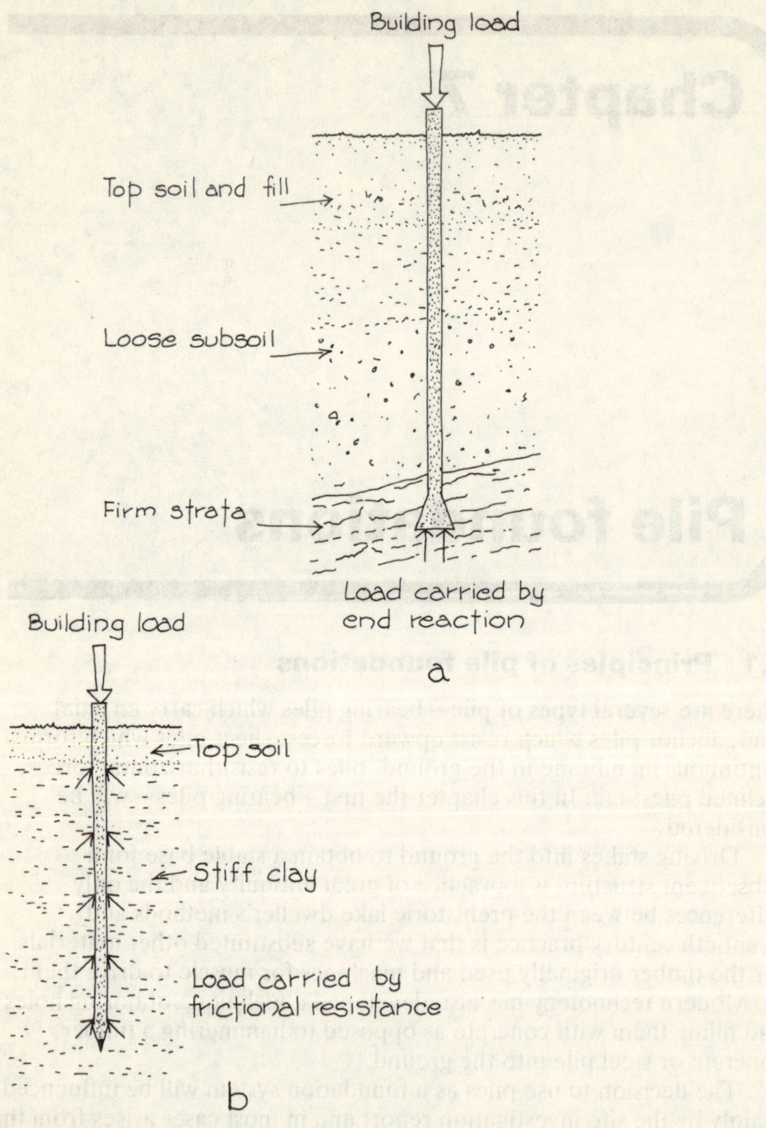

Fig. 7.1 End-bearing and friction piles

resistance between the surface of the pile and the adjoining ground. All piles gain their bearing capacity by a combination of frictional resistance and direct bearing at the tip, but one or other of these predominates to give the classification of 'end-bearing' or 'friction' piles.

7.2 Types of pile

Further classifications in addition to the means by which the pile
achieves its support can be according to the material used, the way it is
employed, the method of installation, etc., but the most significant is
one based on the effect of the presence of the pile on the sub-soil.

When a column of concrete, for instance, is driven down into the
ground the soil must be pushed aside to provide space for the pile. This
movement or displacement of the soil can have considerable
consequences on adjoining buildings, the ground level and the
long-term performance of the pile. As an alternative, the soil can be
removed by boring as already mentioned, and the resultant cavity filled
with wet concrete, usually reinforced, which replaces the soil, thus
avoiding any problems associated with movement of the soil. Figure
7.2 illustrates these two types of pile.

Within the classification of 'displacement' and 'replacement' piles
there are many variants, in the first case mainly concerned with
material and the shape of the pile; and in the second case the material
must be concrete with the variants based on the installation methods.

7.2.1 Displacement piles

The driven piles used in this method are either solid or hollow but
closed at the driving point. The solid piles are either the traditional
timber (still employed in certain conditions), pre-cast concrete or steel,
the most common being concrete.

Timber piles are frequently used in parts of the world where
suitable tree trunks are readily available. They are light, flexible and
resistant to shock, which makes them suitable for temporary work but,
as is the case with any timber post in the ground, they are prone to
decay at the 'wind and water line', i.e. in the area of fluctuating wet
and dryness. If kept permanently wet or permanently dry timber will
last a very long time (the timber work of Roman times has been found
to be in perfect condition where it has been continuously buried in
water-bearing strata). The lower end of a timber pile is fitted with a
cast steel point retained by mild steel straps and the top end is shaped
to a circular section and fitted with a mild steel hoop to prevent
splitting (see Fig. 7.3).

Concrete piles, when pre-cast, have much the same profile as
timber except that a steel or cast iron point is only required where
sub-soil containing cobbles or boulders has to be negotiated. Figure 7.3
shows the details of a typical pre-cast concrete pile and the types of
pile toes.

Reinforcement of the concrete is required firstly to resist stresses
when lifting and handling, secondly to resist stresses during driving and
finally to assist in carrying the building load. Of these three, the last
and permanent use imposes the least stress, the tendency to split or
burst during driving is resisted by the lateral reinforcement and the

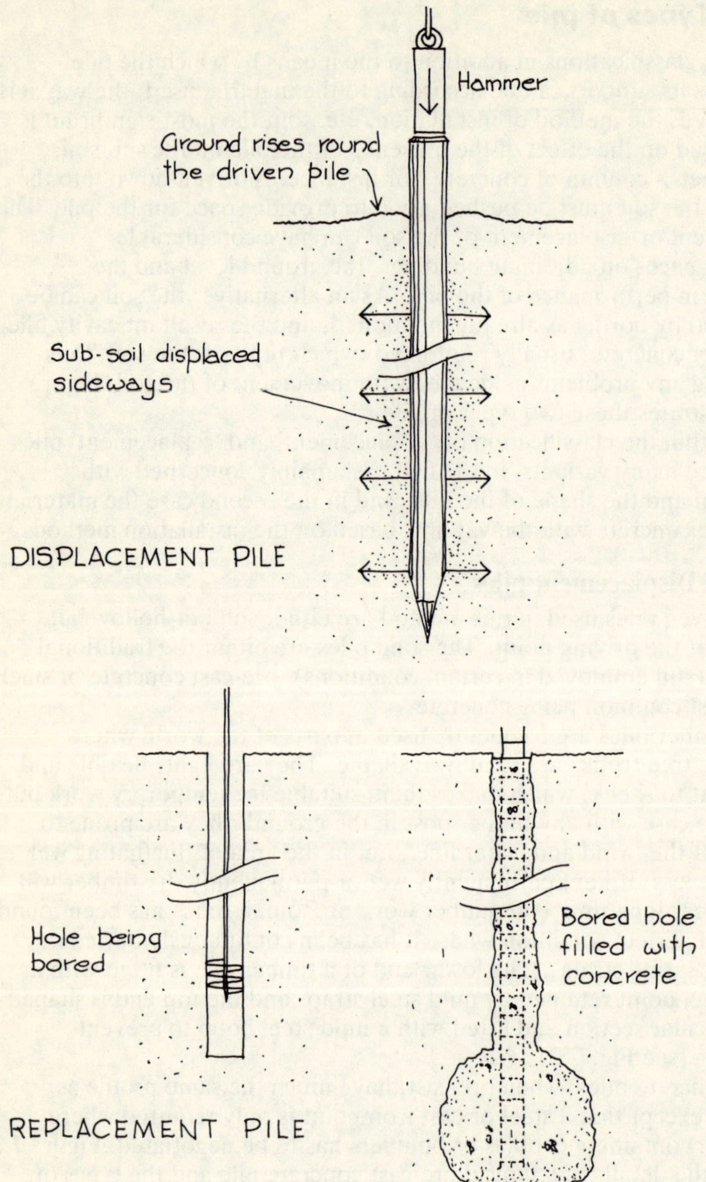

Fig. 7.2 Types of pile

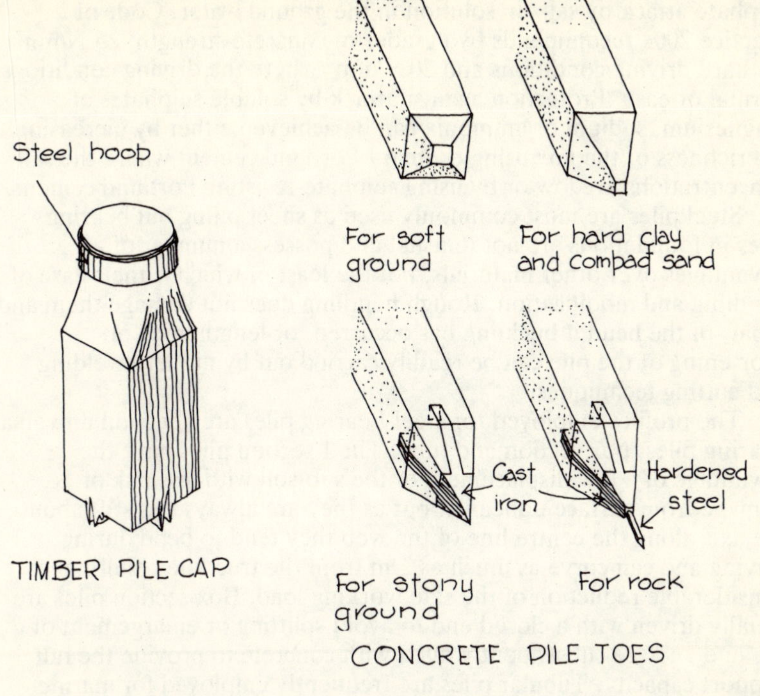

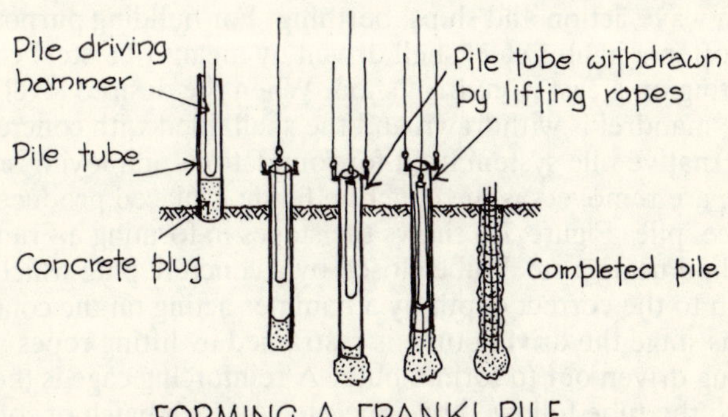

Fig. 7.3 Types of displacement pile

handling moment which occurs when one end of the pile is lifted from a horizontal position is resisted by the longitudinal steel working in the same way as reinforcement in a beam. To assist in hoisting, a short length of steel tube is cast in to facilitate the fixing of toggle bolts, to which the hoisting rope is attached.

In selecting the concrete to be used for piles, regard has to be taken of the working stresses to be encountered and the possibility of sulphate attack by salts in solution in the ground water. Code of Practice 2004 recommends two grades of concrete strength: 25 N/mm^2 for hard driving conditions and 20 N/mm^2 where the driving conditions are normal or easy. Protection against attack by soluble sulphates of magnesium, sodium or ammonia can be achieved either by increasing the richness of the mix using ordinary Portland cement where the salt concentrations are low or by using sulphate-resisting Portland cement.

Steel piles are most commonly used as sheet piling but bearing piles in foundations are not unusual and possess a number of advantages over other materials, not the least of which is their ease of handling and modification. Rough handling does not damage them and repair of the head if buckling has occurred, or lengthening or shortening of the pile can be readily carried out by modern welding and cutting techniques.

The profiles employed for steel-bearing piles are I section universal bearing piles, box section and tube. The I section piles have the advantage of a low displacement of the sub-soil with less risk of damage from surface upheaval, but as they are always less stiff about the axis along the centre line of the web they tend to bend during driving and can curve as much as 2 m from the true line resulting in considerable reduction of the safe working load. Box section piles are usually driven with a closed end to avoid splitting or enlargement of the end and are subsequently filled with concrete to provide the full support capacity. Tubular piles are frequently employed for marine work because the shape has a good resistance to the lateral forces arising from wave action and ships' berthing. For building purposes, the tube is often a lighter steel shell driven by means of a steel core or mandrel acting on a cast iron drive shoe. When the desired level is reached the mandrel is withdrawn and the shell filled with concrete.

An alternative pile system used for foundations employing steel tubes which are removed as the concrete filling is placed produces a 'cast-in-place' pile. Figure 7.3 shows the stages in forming a Franki pile by this method using a steel tube closed by a concrete plug which is driven down to the correct depth by a hammer acting on the concrete plug. At this stage the driving tube is restrained by lifting ropes and the concrete plug driven out to form a bulb. A reinforcing cage is then lowered into the tube followed by concrete. As each batch of concrete is placed it is consolidated by the hammer whilst the tube is being withdrawn and this forces the concrete out into the ground and produces a rough surface to the pile. The process requires skill in its execution but the advantages of increased end bearing because of the enlarged bulb at the bottom and the increased frictional resistance because of the roughness of the surface make this a popular system of displacement piling.

7.2.2 Replacement piles

If a tubular pile is driven without a shoe the amount of displacement is restricted to the negligible volume of the steel in the tube. Subsequent removal of the soil within the tube and filling with concrete produces one type of replacement pile. The other and more common type is a bored pile where the soil is removed by an auger or a grab, the bore hole being unlined or lined with a steel shell, and the resulting tubular hole fitted with a reinforcing cage and concrete. Figure 7.4 shows details of bored piles of which the simplest is the hand-augered pile.

Simple piles of this sort, limited in size to 350 mm diam. and 4.5 m depth, have a wide application for lightly loaded buildings and when used in conjunction with a reinforced concrete ground beam can be economically competitive with strip foundations for domestic and light industrial work. Since the hole is unlined, this method can only be used

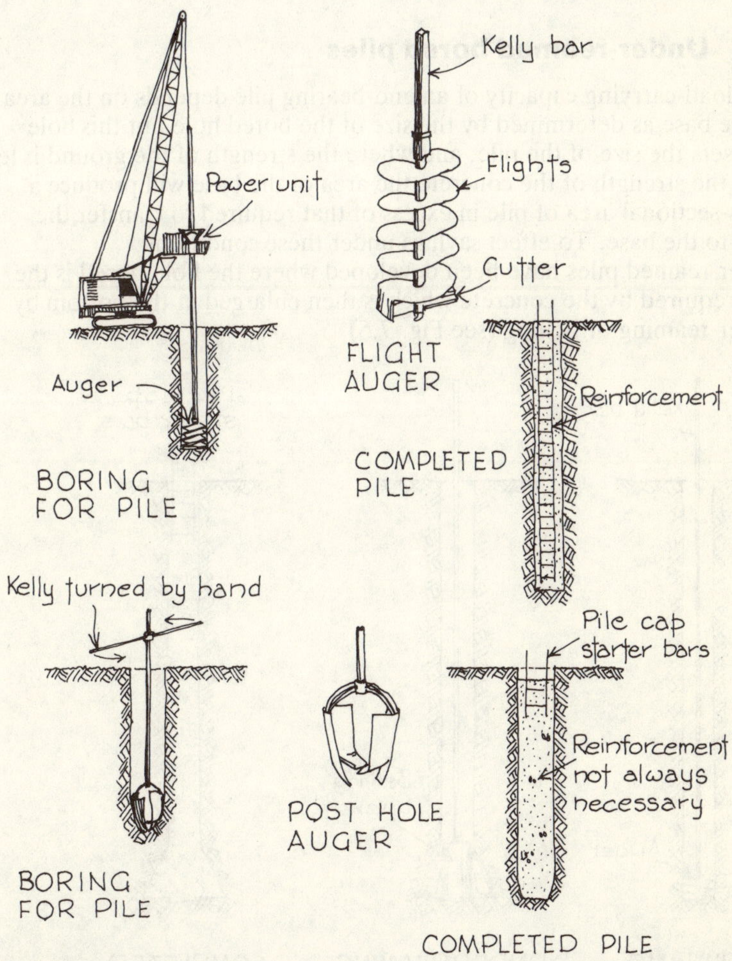

Fig. 7.4 Bored piles

in cohesive soils which are self-supporting. A mechanical auger can drill to depths of 45 m and diam. up to 2 m but, as in the case of hand-augered holes, the ground must be self-supporting and preferably free from cobbles, boulders, etc. Weak soils can be supported by a lining inserted after the auger has been removed or by filling the hole with bentonite slurry.

Difficult soils such as soft clay or silt, stoney clay, or water-bearing gravel require a percussion or grab type of boring in which the excavating head operates within a casing which follows the cutter down (see Fig. 7.4). The percussion boring head is suitable for weak uniform soils but where cobbles, boulders or shale has to be bored, a grab type of boring head is more efficient.

7.3 Under-reamed bored piles

The load-carrying capacity of an end-bearing pile depends on the area of the base as determined by the size of the bored hole but this hole also sets the size of the pile, and where the strength of the ground is less than the strength of the concrete the area of the base will produce a cross-sectional area of pile in excess of that required to transfer the load to the base. To effect savings under these conditions under-reamed piles have been developed where the hole bored is the size required by the concrete which is then enlarged at the bottom by under-reaming or belling (see Fig. 7.5).

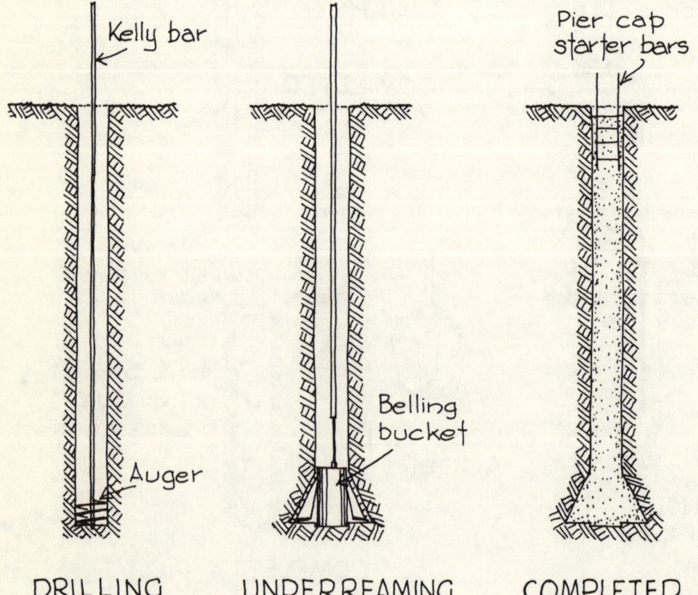

Fig. 7.5 Underreamed pile

If the bore hole is large enough to contain a cage holding men then this under-reaming can be carried out by hand, but in smaller installations the under-reaming must be cut by a belling bucket which is attached to the drilling rods in place of the cutting head and consists of a cylindrical bucket to which two cutting arms are hinged. The cutting arms are pushed outwards by the drill rods to enlarge the bore hole and the loosened material falls into the cylindrical bucket.

Under-reamed bases of up to $2\frac{1}{2}-3$ times the shaft diameter are possible by this method with a maximum diameter of approx. 7 m.

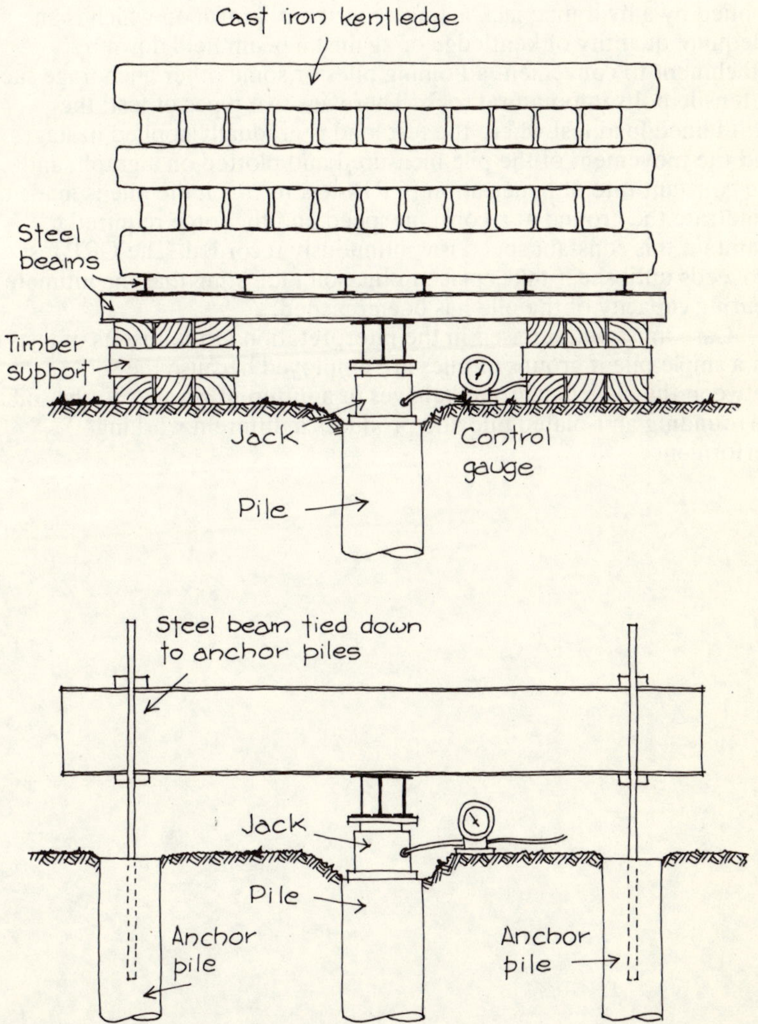

Fig. 7.6 Load tests on piles

7.4 Testing piles

It is recommended that at least one test pile is formed on any piling contract in such a position that it will behave in exactly the same way as the bearing piles but yet not get in the way of the work. The test pile must not be incorporated in the finished building. It is bored to the same depth or driven to the same set (the amount the pile moves per blow) and overloaded to a stipulated amount beyond the design load or to a point of failure.

The method of testing as shown in Fig. 7.6 is to impose a load onto the pile for 24 hours and record the movement, if any. Loading is applied by a hydraulic jack acting against a platform on which is an adequate quantity of kentledge or against a beam held down by attachment to convenient adjoining piles or some other anchorage such as tensile bolts into natural rock. There are two types of test: the maintained load test where the test load is gradually applied in stages and the movement of the pile measured and plotted on a graph, and the constant rate of penetration (CRP) test in which the pile is made to penetrate the ground at a constant speed and the force required to maintain this constant speed is continuously recorded. The CRP test proceeds until shear takes place in the soil indicating that the ultimate bearing capacity of the pile has been reached.

Care must be exercised in the interpretation of the results of tests on a single pile if groups of piles are employed because the soil between the piles in a group behaves in a different manner to the soil surrounding an isolated pile and produces a different working performance.

Chapter 8

High retaining walls

8.1 Function

The fundamental purpose of all retaining walls is to hold back a
material along a vertical face. The detailed function is, therefore,
concerned with the type and quantity (as determined by the retained
height) of the material. In building work, retaining walls are required
to support differences in level of soil; in other applications they may be
required to contain liquids, especially water, or to confine granular
materials to be stored.

Differences in surface levels can be caused by three site activities,
first, where a cut has to be made into a sloping site (Fig. 8.1*a*); second,
where a sloping site is levelled by a process of cut-and-fill (Fig. 8.1*b*);
and third, where the ground level has been lowered for a basement or
tunnel (Fig. 8.1*c*).

8.2 Forces in retained material

If, say, dry sand is deposited on a level surface it will form a heap with
a surface at about 30° to the horizontal. Any attempt to increase the
height with more material will result in the sand running down the
heap until the base size has increased sufficiently to restore the 30°
slope. This is the angle of repose or the angle of shear and varies
according to the material, its granular properties and its condition.

Whatever the angle is, a bank finished at this angle between two

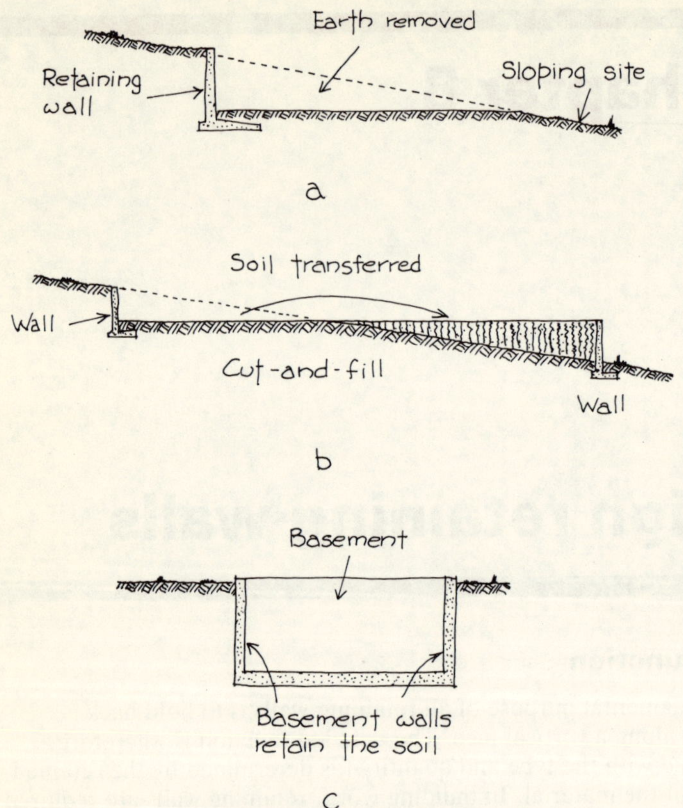

Fig. 8.1 Where retaining walls are needed

levels will not require a retaining wall no matter how high because the soil has no tendency to move. It therefore follows that the forces exerted on a retaining wall are solely due to the material contained between the angle of repose and the back of the wall (see Fig. 8.2*a*), and of this material half rests on the soil below and only half bears on the wall.

This may not be all that the wall has to do: if a load is placed on or in the wedge of soil above the angle of repose it will tend to force it along the angle and thrust against the back of the wall. This additional load, or surcharge, can be due either to foundations within the area or to surface features such as traffic on a road or material piled on the upper level as shown in Fig. 8.2*b*.

One of the factors which affects the angle of repose and hence the pressure on the wall is the presence of moisture. Reference has just been made to dry sand, and it is common knowledge that if the sand is damp it will stand at a much steeper angle. However, if the sand is

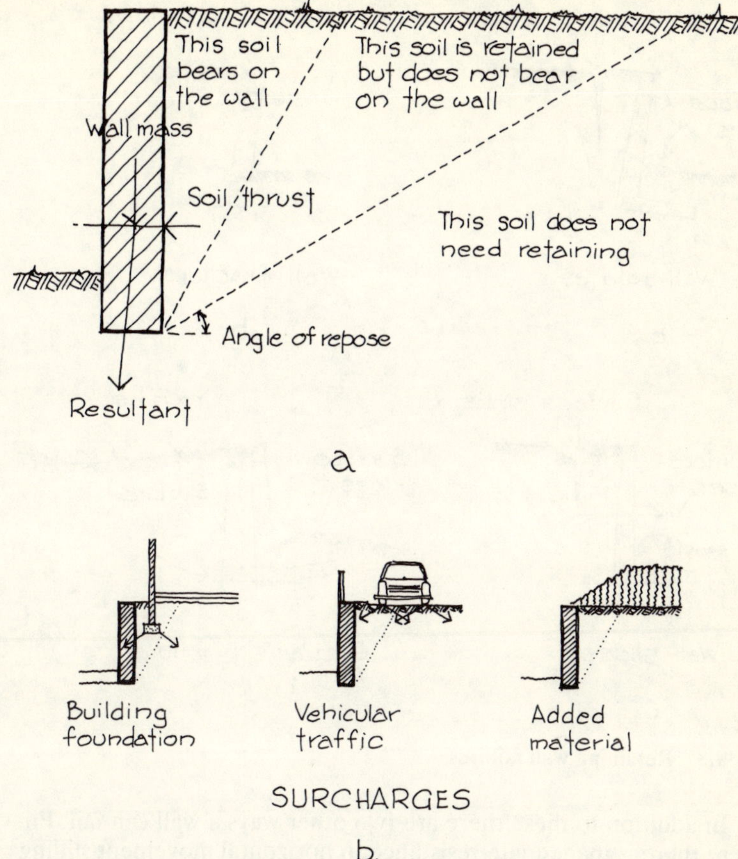

Fig. 8.2 Retaining wall loads

made so wet that the grains are separated by water, i.e. it is saturated, the mass acts like a fluid and the angle of repose is virtually nil. Thus the forces acting on the wall would become horizontal and cause a more severe stressing.

8.3 Failure of retaining walls

The form of failure which most obviously must be avoided is the inability of the wall to counteract the thrust and, as a result, overturning. This inability can be due either to a fault in the design or to overstressing of the soil under the base of the wall (see Fig. 8.3a). Faulty design can also result in the structure of the wall itself giving way as shown in Fig. 8.3b.

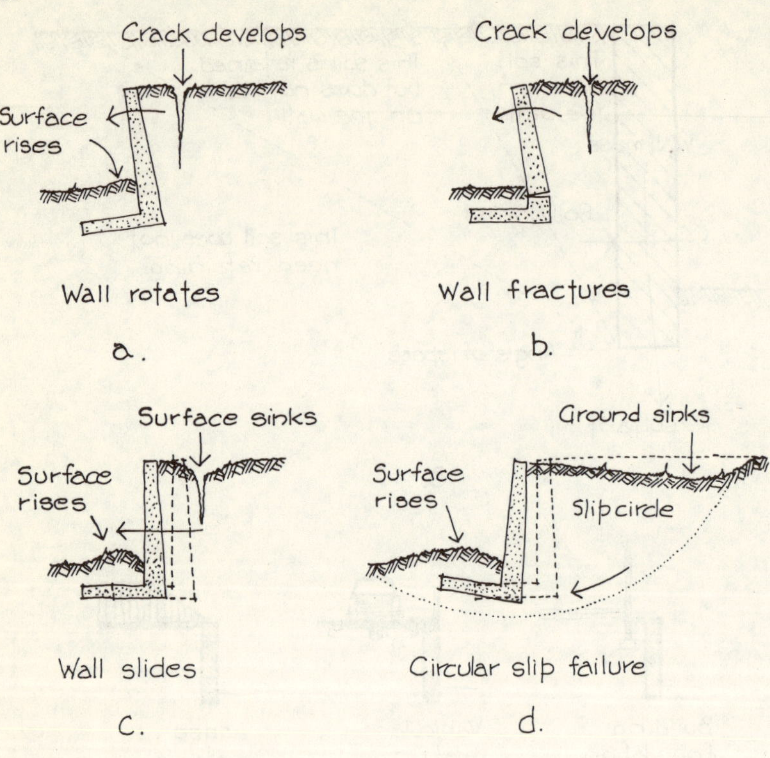

Fig. 8.3 Retaining wall failures

In addition to these there are two other ways a wall can fail. First, where there is inadequate resistance to horizontal movement sliding can occur (Fig. 8.3*c*); or second, in low-strength cohesive soils the difference in mass of soil each side of the wall can cause the ground to move along a curved slip plane carrying the wall with it as in Fig. 8.3*d*.

8.4 Types of retaining wall

There are two main types of retaining wall – those which work by balancing the mass of the wall against the thrust of the retained material, which are known as gravity-retaining walls (Fig. 8.4*a*), and those which resist the forces by their strength as a vertical cantilever (Fig. 8.4*b*) or as a horizontal slab on edge between supports (Fig. 8.4*c*).

A gravity wall must be of sufficient mass to bring the line of action of the resultant between the wall and the imposed force within the middle third of the base, as shown. As a guide to this the base thickness should be about one-third the retained height and the top thickness about one-seventh the height. For this reason a gravity wall is

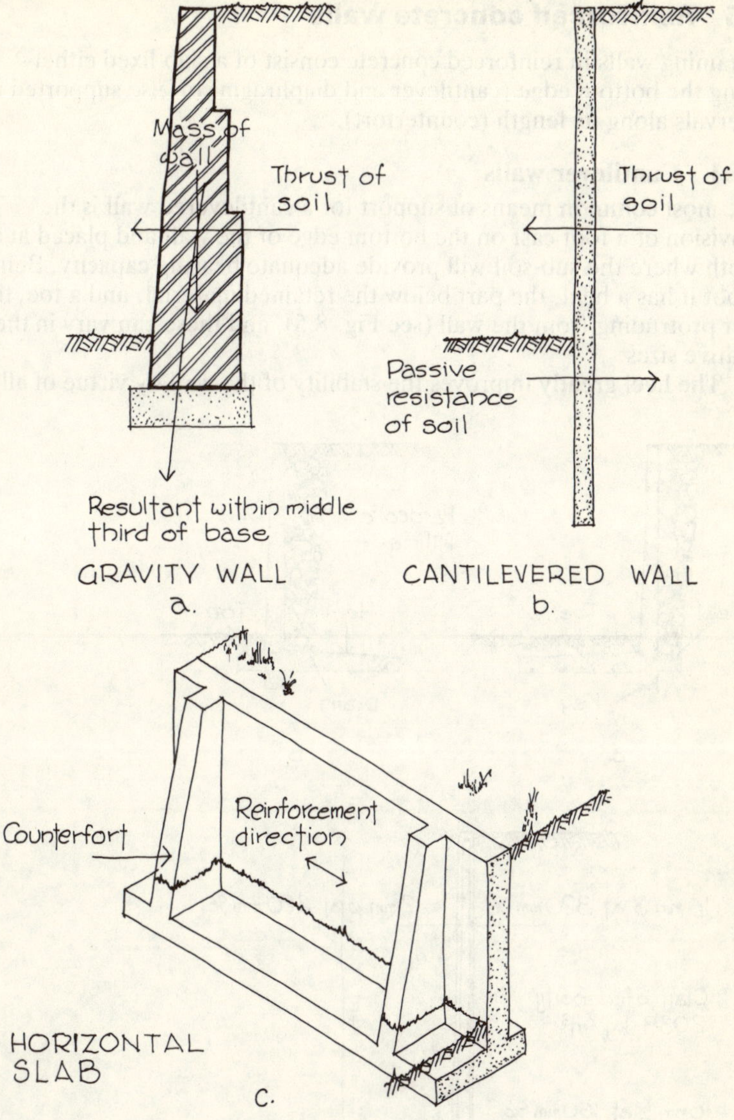

Fig. 8.4 Types of retaining wall

not an economic proposition if the height exceeds 2 m, not only because of the quantity of material used in its construction but also because of the area it takes up on the site.

High retaining walls, in this case up to 6 m, are always one of the forms which rely on strength and are either reinforced concrete or steel.

8.5 Reinforced concrete walls

Retaining walls in reinforced concrete consist of a slab fixed either along the bottom edge (cantilever and diaphragm) or else supported at intervals along its length (counterfort).

8.5.1 Cantilever walls

The most common means of support for a cantilevered wall is the provision of a foot cast on the bottom edge of the wall and placed at a depth where the sub-soil will provide adequate bearing capacity. Being a foot it has a heel, the part below the retained material, and a toe, the part protruding from the wall (see Fig. 8.5), and these can vary in their relative sizes.

The heel greatly improves the stability of the wall by virtue of all

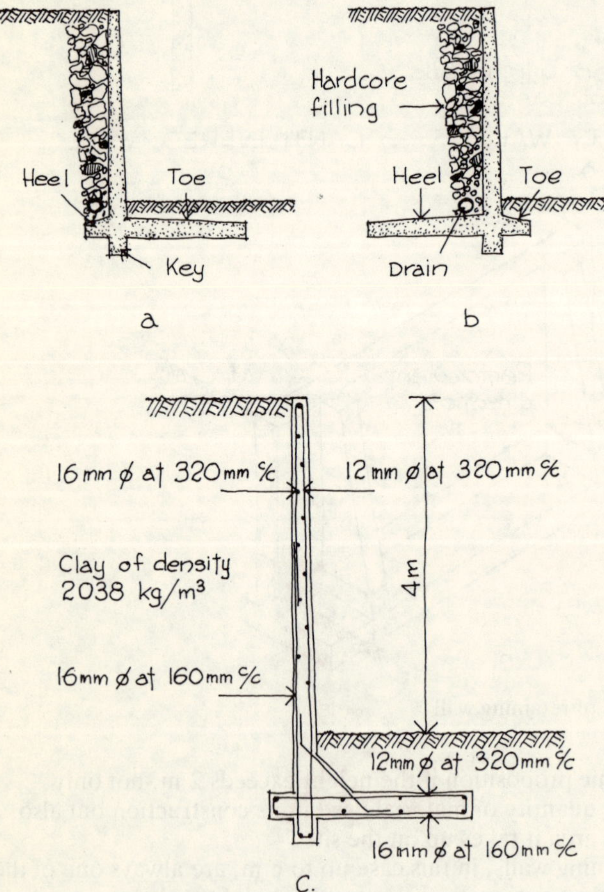

Fig. 8.5 Cantilever retaining wall

the soil resting on it but to form a large heel it is necessary to excavate well back into the upper level; an operation which not only will increase the costs but may be impossible if, say, the retaining wall is to be along a boundary and would involve excavating in the adjoining property. The toe is less efficient in its support but is far easier to form and thus the most common version of this type of wall is as shown in Fig. 8.5c. The key shown projecting down from the base is to increase the passive resistance to horizontal sliding. The reinforcement only relates to a wall retaining a height of 4 m of clay of 2038 kg/m^3 density and will be different in other cases.

An important feature is the provision of hardcore and a land drain or weep holes behind the wall. Their purpose is to prevent the build-up of ground water and the creation of a hydrostatic pressure capable of overturning the wall.

The form of wall shown in Fig. 8.5c is available in pre-cast units up to 3.6 m high. These units are set on a concrete base formed *in situ* and are connected by a joggled joint down their meeting edges. They can be used as permanent walls or as temporary enclosures for the storage of loose materials.

8.5.2 Diaphragm walls

This type is more simple than the last; it is a true cantilevered slab but it is generally more difficult to form. It comprises a continuous sheet of reinforced concrete carried sufficiently far below the lower level to provide an anchorage for the exposed upper section, as shown in Fig. 8.6. The method is attractive when the difference in ground levels is created by excavation since the diaphragm can be constructed in advance of the reduction of ground level, thus eliminating any need for temporary support.

There are two methods of construction: either a row of interlocked bored piles – 'contiguous' or 'secant' piling – is formed or else a series of wall panels are cast in a predetermined order in a narrow trench to form a continuous wall.

The contiguous piling method (Fig. 8.6a) comprises boring and casting a row of piles to the requisite depth below the excavated level at a distance apart equal to one and a half diameters of the piles. Before the concrete of these has reached working strength, a second set of borings are made between the first which removes both the intervening soil and some of the 'green' concrete of the first set of piles, thus forming a recess into which the second set of piles lock to produce a continuous structure.

The wall panels method (Fig. 8.6b) employs the slurry-filled trench principle whereby, after forming guide walls down to about 1 m, a short length of trench is taken out by grabs or rotary drilling. To prevent the sides of the trench collapsing, it is continuously filled with bentonite slurry as excavation proceeds. Bentonite slurry is a thixotropic suspension – a gel which becomes a liquid when disturbed –

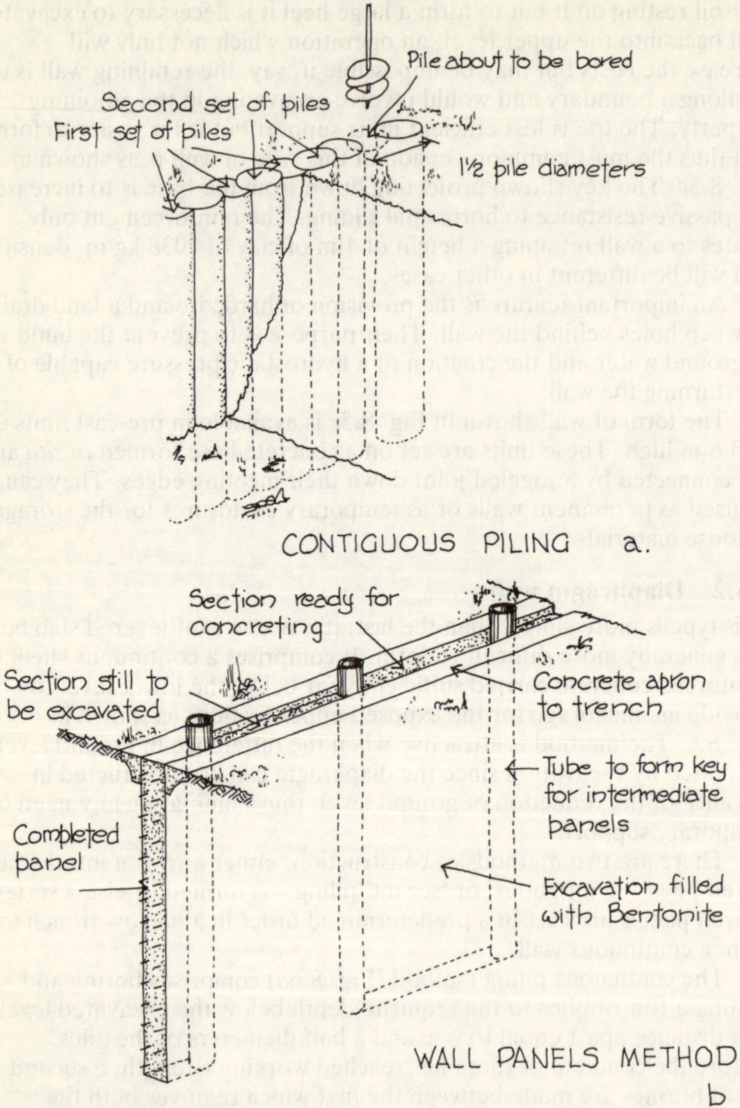

Fig. 8.6 Diaphragm walls

and thus will support the trench walls but does not interfere with construction operations. Once the section of trench has been excavated a tube with an outside diameter equal to the trench width is placed at each end to form a semi-circular recess into which the next panel will interlock and a reinforcement cage lowered in between the tubes. Filling of the trench with concrete is effected through a tube called a tremie pipe and, as the trench fills with concrete, the displaced bentonite slurry is pumped either into a tank or directly into an adjoining section of trench being excavated. When the first panels have reached working strength the end tubes are withdrawn and intermediate panels are excavated and filled by the same process to form a continuous structure.

An advantage of the second method over the piled method is that the face exposed by excavation after the diaphragm has been formed is relatively flat and true and requires less finishing than the surface presented by contiguous piling.

8.5.3   Counterfort walls

In section a counterfort wall is similar to a cantilever wall but because the support is provided by the buttressing of the counterforts, the slab is designed to span horizontally rather than vertically (Fig. 8.4c).

The counterforts may be constructed on the upper or lower side of the wall depending on circumstances and requirements. In the first case the upper level must be cut back to allow for the construction at the counterforts and thus the wall cannot be along a boundary, though this does present a clear face to the lower level; in the second case no problem exists with construction but the counterforts can be expensive to finish neatly and form an obstruction of the lower area.

As the cost of forming the counterforts is high, the method only becomes economical for retaining walls over 5.0 m high or where a high surcharge is imposed, when the saving in thickness over a cantilevered wall offsets the cost of the counterforts.

A similar system arises with basement walls, which are considered as retaining walls since they must resist the lateral pressure of the adjacent soil except that the buttressing is provided by the end return walls and any intermediate partition walls.

8.6   Steel retaining walls

The usual method adopted for steel walls is the true cantilever as described in Section 8.5.2 whereby the support for the retaining part of the wall is obtained by embedding it into the ground below the lower level.

In this case sheet steel piling would be used (Fig. 8.7) driven down to the required depth in relation to the angle of repose of the soil; and, as can be seen, for a 30° angle of repose (a fairly common angle) the length of the sheet piles has to be more than twice the retained height. This length can be reduced considerably if the top of the piling can be retained by ground anchors as shown in Fig. 8.7b.

8.7   Other methods of retaining soil

Where space allows or particular conditions are to be met such as tidal

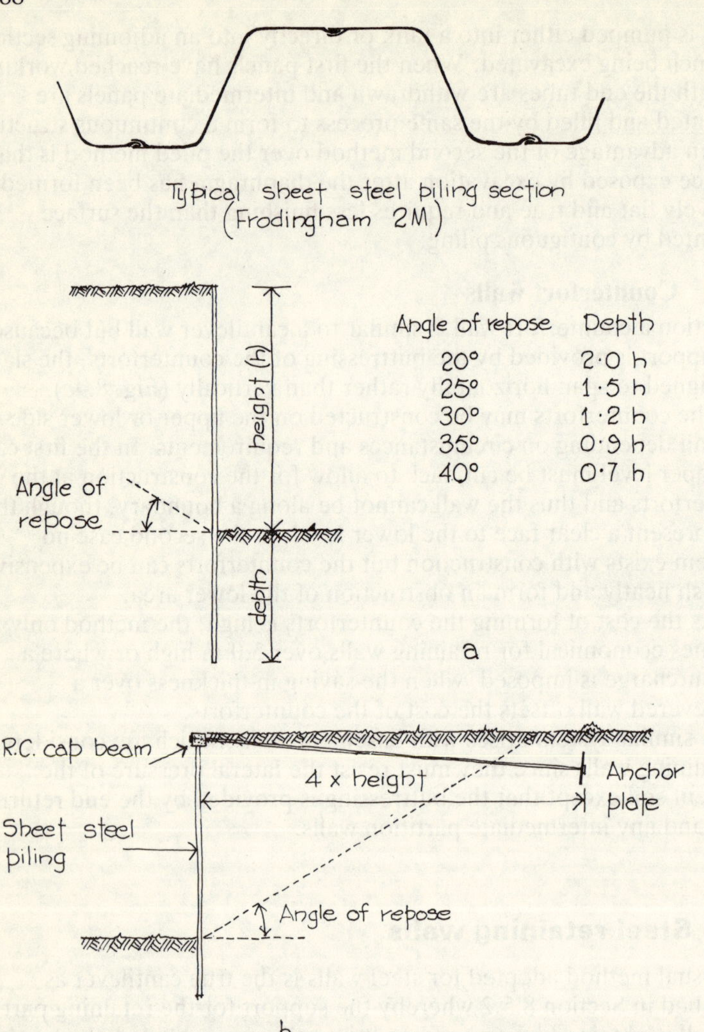

Fig. 8.7   Sheet steel piling

and wave effects along the coast, the transition from higher to lower level may be by a bank at, or very near to, the angle of repose. To give this a permanence, particularly in coastal and reservoir work where this is often found, a line of sheet steel piling is driven in along the bottom edge of the slope to prevent undercutting, and the slope is faced with dressed stone or concrete blocks. The completed work is known as a revetment (see Fig. 8.8*a*).

Another method gaining popularity because of its simplicity and low cost is the concrete crib wall (Fig. 8.8*b*). This is really a gravity wall system comprising pre-cast concrete H-shaped cross members and

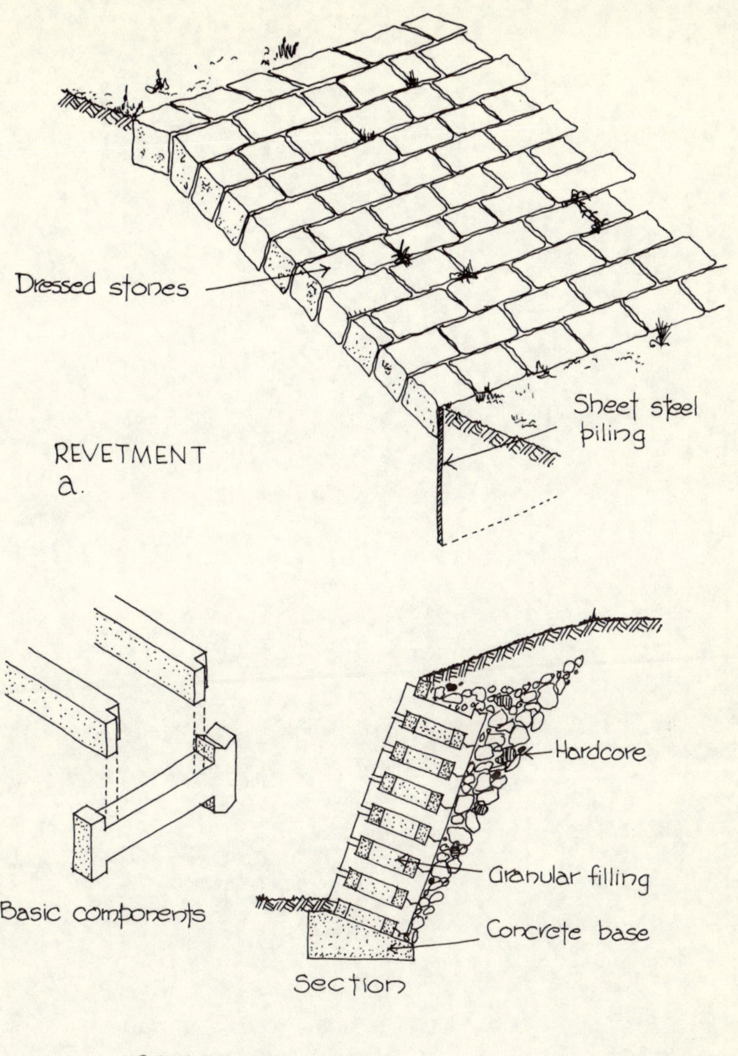

Fig. 8.8 Alternative retaining methods

horizontal rectangular beams. These are interlocked and the
intervening spaces filled with soil or broken rock. To improve its
stability it is usually built to a slope or batter off an *in situ* concrete
base and the concrete units are laid dry. It gains its economic
advantage because the largest part of the wall is the granular fill which
is usually readily available on the site.

Part III

Superstructure

Chapter 9

Complex scaffolding

9.1 The need for complex scaffolding

The normal putlog or independent forms of scaffolding will satisfy the great majority of demands for access for building purposes, but occasionally a different demand arises or a condition exists which make the normal form of scaffold inappropriate or impossible.

One such demand is where the work to be done is at high level and over a large area; quite often this would be an internal building operation such as ceiling work. To provide the working area required in this case either a complete platform on a two-dimensional structure, called a bird-cage scaffold, is erected or, where the work is of a short duration at any position, a mobile tower platform should be used.

Work of short duration on a vertical face also means that a mobile working platform is required, in this case moving both horizontally and vertically, usually slung from a structure at the top of the building. This type also provides a working platform which is readily adjustable to a precise level.

The most common reason why a normal scaffold cannot be erected is that support at ground level is not available or is restricted, either because of the obstruction that the scaffold standards would present or else because the working platform is at such a level that the height to be covered makes it uneconomic. In these cases the scaffold frame is cantilevered from the building, independent of support from the ground and either hung from the top, 'suspended', or from the bottom, 'truss-out'.

9.2   Bird-cage scaffold

As already mentioned, this type of scaffold presents a large working area for operations at high level but because of its cost it is only feasible in cases where the work is going to last a long time. An alternative use for a bird-cage scaffold is to provide support to a structure such as a floor slab or bridge deck. For this purpose the framework would be carefully designed and calculated according to the magnitude, nature and direction of loading, the height of the structure and the bearing capacity of the surface on which the scaffold will stand. This alternative use is generally described as falsework.

An access bird-cage scaffold would be constructed of standards (upright members), ledgers (horizontal members) and transomes (platform bearers) all in steel tube and connected by couplings in the same manner as a normal independent scaffold. In most cases bracing will also be required to achieve working stability or to resist wind loading.

The maximum spacing for the standards is 2400 mm and may need to be less than this if any degree of dead loading is to be anticipated. Using normal 50 mm (nominal) steel tube the safe uniformly distributed load (UDL) on a bird-cage scaffold with standards of 2400 mm centres in each direction and a lift height of 1800 mm would be approx. 5.5 kN/m^2, whereas reducing the standard spacing to 1500 mm increases the safe UDL to approx 13.5 kN/m^2.

In common with all scaffolding structures, the working platform, its support and the necessary safety measures must all be carried out in accordance with BS 1139 pts. 1, 2 and 4: 1982 and also The Construction (Working Places) Regulations 1966.

9.3   Mobile tower scaffold

Mobile towers are frequently used by painters, electricians, maintenance crews and any other workmen who need to gain access to overhead areas for relatively short periods of time at any one point.

The construction is either of standard scaffold tube and couplings with purpose-made wheels or of a patented system of framed units as described in section 9.1, but whichever method is used it must comply with Construction (Working Places) Regulations 1966. In plan the tower is square and has a working platform at the top with a built-in ladder or stairway access (see Fig. 9.7). The working area must be provided with guard rails and toe boards as for a normal static scaffold and the wheels or castors must be capable of being locked so that the wheel cannot turn and the mounting cannot rotate in the bottom of the tube.

To prevent overturning in use the ratio of height to plan dimension must be not less than 3.5 to 1 for towers up to 4.5 m high. Over this

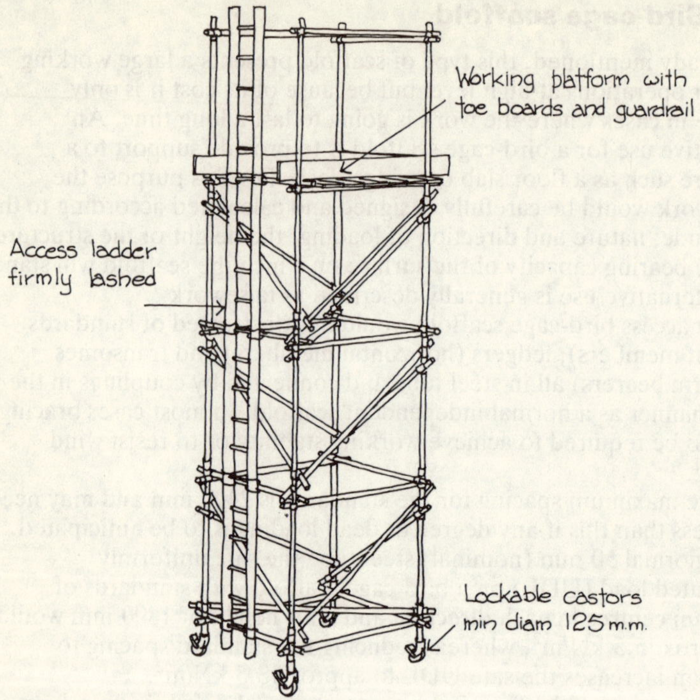

Fig. 9.1 Mobile tower scaffold

height the size should be calculated. When used outside they should always be stabilised with guy ropes; internally this is not usually necessary provided the floor is smooth and level, if the height is below 4.5 m and no horizontal loads need be applied.

One cause of overturning of towers is the practice of climbing up the outside instead of up an internal ladder. By so doing the workman is relying on the weight of the tower to counterbalance his own weight and in some cases, particularly small aluminium tube towers, this is not enough.

In use it is prudent to check the tower carefully before climbing to make sure that all four castors are fully supported (adjustable ones are available to overcome any problems of uneven surfaces), that they are locked solid and that the whole structure is properly stabilised with ropes where necessary.

9.4 Slung scaffold

Another means of working at high level internally is by the use of tubes fixed to wire ropes attached to suspension points in the roof framing of the building. A working platform is constructed across the tubes and the working space properly protected by guard rails and toe boards, and the whole structure is secured by some suitable means to prevent undue horizontal movement.

The Construction (Working Places) Regulations 1966 define a slung scaffold as 'a scaffold supported by means of lifting gear, ropes or chains or rigid members and not provided with means of raising or lowering'. The absence of the ability to raise or lower it is the main difference between this and a suspended scaffold. The Regulation also lays down that if rope is used it must be wire rope.

9.5 Cantilevered scaffold

This, as already explained, is the term applied to a scaffold frame which obtains its support from the building by a lattice framework attached to and bearing on the structure. Figure 9.2*a* shows a typical arrangement of a 'truss-out' cantilevered scaffold. The Construction (Working Places) Regulations 1966 also refer to this form as a Jib, Figure and Bracket scaffold.

It is obvious that great care must be exercised in the design of this scaffold and particularly in the choice and detail of anchorage to the building. Code of Practice 97 Part 3 recommends jacking a frame between structural floors to obtain a frictional resistance to the horizontal forces. Where possible a more positive anchorage with bolts or scaffold tube set into the structure will achieve a more secure scaffold.

A more direct approach to the design of a cantilevered scaffold is shown in Fig. 9.2*b* where the truss or jib frame is replaced by a rolled steel beam attached to the floor and projecting from the building as a true cantilever. The tailing-down of the cantilever beam must be either by means of bolts passing through the floor as shown or by the use of a thrust tube framework between the top flange of the beam and the underside of the floor above. Packs of suitable material are provided at the fulcrum point of the beam so as to control the position of the point of bearing. The scaffold frame itself is braced back to the beam and should also be tied to the floors or walls above to achieve stability.

As the thrusts and pulls and forces in the members of the scaffold frame can be considerable, a cantilevered scaffold should always be designed and calculated by a qualified engineer and installed with great care, especially if it is to be used for hoisting and loading materials as well as access.

9.6 Suspended scaffold

A suspended scaffold is also a cantilevered structure but in this case the

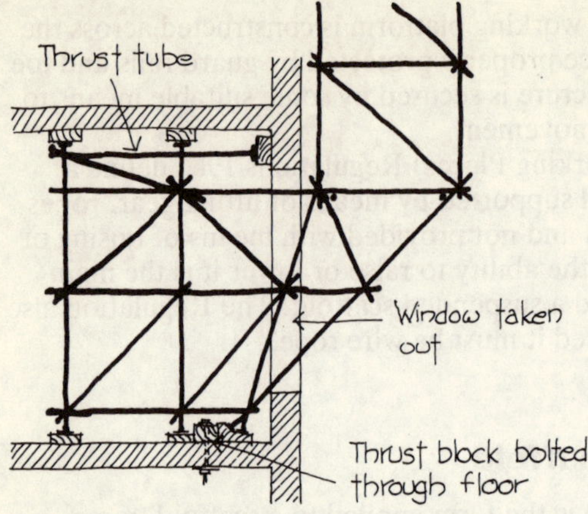

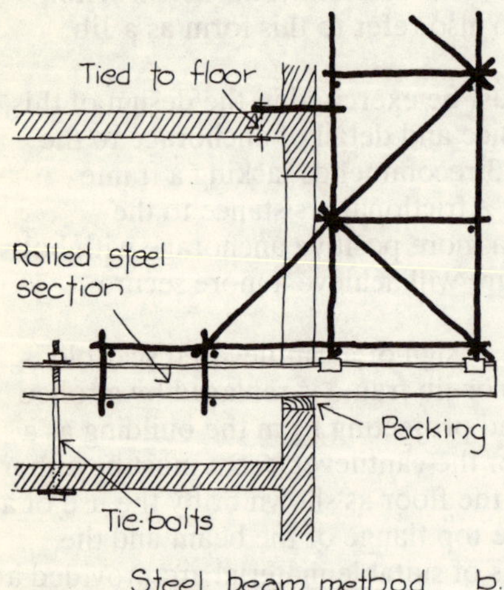

Fig. 9.2 Cantilevered scaffold

cantilever is at the top and the working platform hung below (see Fig. 9.3). The cantilever beam or outrigger must be securely supported and fixed or held down with counterweights which, according to the Construction (Working Places) Regulations 1966, must be not less than three times the weight which would counterbalance the total weight suspended from the outrigger at any time.

The Regulations also require that the suspension is by wire, rope or chains and of such a length that when the platform is at its lowest

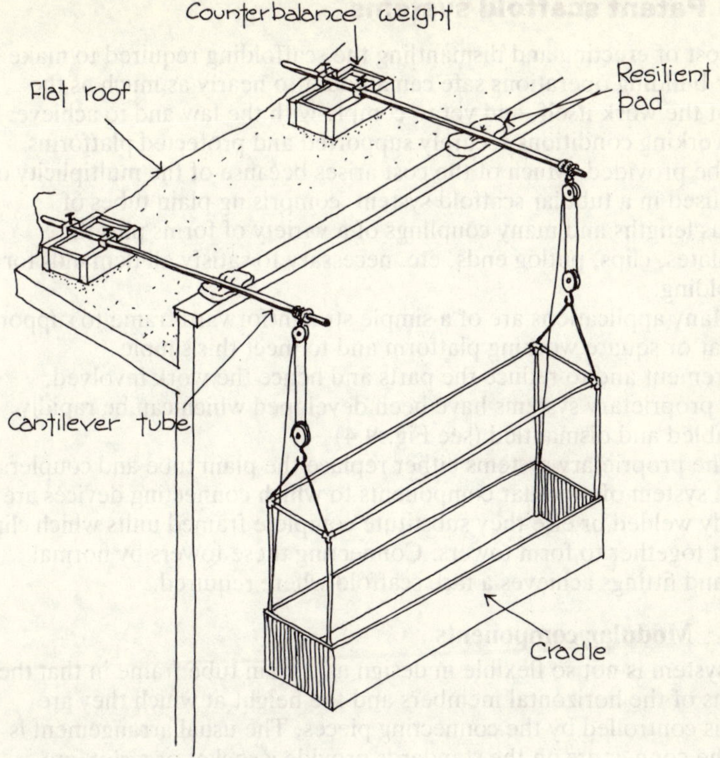

Fig. 9.3 Suspended scaffold

level at least two turns are left on the winch drum.

The working platform can vary from a one-man bosun's chair or a two-man cradle or gondola to a continuous articulated platform the length of the building suspended at a series of points. Whatever its size or form it must be closely boarded and of adequate width for working, subject to a minimum of 25 inches (635 mm) when used only for footing and 34 inches (863.6 mm) if used for the deposit of materials (Regulation 19 (13)). It must also be so arranged as to prevent undue tipping, tilting or swinging, secured so as to prevent undue horizontal movement, and as near to the building as practicable.

As this form of scaffold can be raised and lowered at will it provides ready access to the outside face of buildings for maintenance purposes. For this reason many tall buildings with flat roofs have permanent tracks fixed around the perimeter of the roof on which a trolley runs bearing davits which can be swung out over the building face to support a suspended cradle. The winches are operated from the cradle giving the workman full control over his working position.

9.7 Patent scaffold systems

The cost of erecting and dismantling the scaffolding required to make minor building operations safe can amount to nearly as much as the cost of the work itself, and yet to comply with the law and to achieve safe working conditions securely supported and protected platforms must be provided. Much of the cost arises because of the multiplicity of parts used in a tubular scaffold system, comprising plain tubes of various lengths and many couplings of a variety of forms plus the baseplates, clips, putlog ends, etc. necessary to satisfy all demands for scaffolding.

Many applications are of a simple straightforward frame to support a linear or square working platform and to meet this simple requirement and to reduce the parts and hence the work involved, many proprietary systems have been developed which can be rapidly assembled and dismantled (see Fig. 9.4).

The proprietary systems either replace the plain tube and couplers with a system of modular components to which connecting devices are already welded or else they substitute complete framed units which clip or slot together to form towers. Connecting these towers by normal tube and fittings achieves a face scaffold where required.

9.7.1 Modular components

This system is not so flexible in design as a plain tube frame in that the lengths of the horizontal members and the height at which they are fixed is controlled by the connecting pieces. The usual arrangement is that the connectors on the standards provide a socket or a slot into which a corresponding pin or lug on the end of the ledgers and transomes is dropped and secured with a locking wedge or screw (see Fig. 9.4*a*). It is a useful system for small scaffolding jobs where the plan shape is complex but rectangular and the requirements of platform height are variable.

9.7.2 Frame and brace system

There are two basic components in this system: a main tubular frame and a cross brace frame. The main frames fit into one another by spigots and sockets and the braces are drilled to fit studs on the main frame onto which they are secured by a quick-locking device, as in Fig. 9.4*b*.

The system repeats vertically and longitudinally to form a continuous structure but does not repeat laterally and therefore generally requires the use of normal tube and fittings for tying-in to the building. It can also be used to construct access towers.

9.7.3 The 'H' frame system

With this there is only one basic unit, an 'H' shaped welded tube frame. The vertical members are fitted with spigots to receive the next

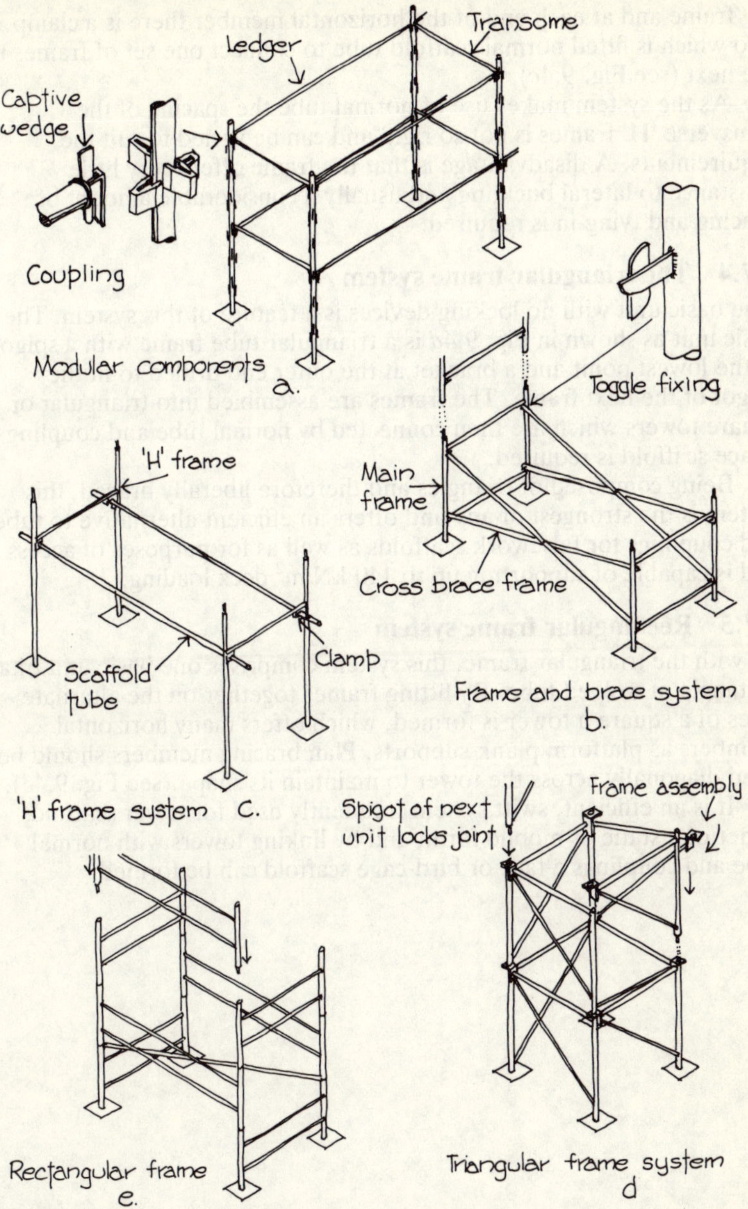

Fig. 9.4 Patent scaffold systems

100

'H' frame and at each end of the horizontal member there is a clamp into which is fitted normal scaffold tube to connect one set of frames to the next (see Fig. 9.4*c*).

As the system makes use of normal tube the spacing of the transverse 'H' frames is not so rigid and can be varied to suit the requirements. A disadvantage is that the frame offers very little resistance to lateral buckling and usually a considerable amount of bracing and tying-in is required.

9.7.4   The triangular frame system

One basic unit with no locking devices is a feature of this system. The basic unit as shown in Fig. 9.4*d* is a triangular tube frame with a spigot at the lowest point and a bracket at the outer end drilled to fit the spigot of the next frame. The frames are assembled into triangular or square towers which are then connected by normal tube and coupling if a face scaffold is required.

Being composed of triangles and therefore liberally braced, this system is the strongest of any and offers an efficient alternative to tube and couplings for falsework scaffolds as well as for purposes of access and is capable of supporting up to 140 kN/m^2 deck loading.

9.7.5   Rectangular frame system

As with the triangular frame, this system comprises one basic unit of a rectangle of welded tube. By fitting frames together on the alternate sides of a square a tower is formed, which offers many horizontal members as platform plank supports. Plan bracing members should be fixed diagonally across the tower to maintain its shape (see Fig. 9.4*e*).

It is an efficient, swift system frequently used for tower scaffolds either of a static or mobile form, and by linking towers with normal tube and couplings a face or bird-cage scaffold can be formed.

Chapter 10

Laminated timber

10.1 Laminated construction

The modern practice is to use glued-laminated timber which the
development of modern adhesives has made possible, but work has
been done in the past using bolted, nailed or pegged laminated
members. These are not so efficient as present methods but
nonetheless produced advantages over solid timber members. Early
examples of this are a number of laminated timber railway bridge
arches constructed between 1830 and 1850 using oak pegs or trenails to
connect the laminates. The development of casein glue at the the turn
of the century led to the first patents for glulam beams being taken out
about 1909 by Hertzer. Many of these structures are still standing.

British Standard 4169 : 1970 currently covers *Glued Laminated
Timber Structural Members* but is being revised and will be replaced.
The Standard defines these members as 'structural components
manufactured from separate pieces of timber with the grain parallel to
the axis of the member, and glued together to form a member which
functions as a single structural unit'. The maximum thickness of timber
should not exceed 45 mm according to the Standard (and is often less
depending on the shape to be formed) and the adhesive should be
casein, urea-formaldehyde, resorcinol or phenolic type; animal glues or
vinyl, epoxy resin or rubber-based adhesives are not suitable.

Laminated timber is used as a straight or slightly cambered beam
to achieve economic and structural advantages over a solid timber

member and as a curved member to make use of the great structural strength these can attain or to fulfil a design requirement.

10.2 Materials

10.2.1 Timber

The BS states that the timber can be any of the softwoods or the hardwoods set out in CP 112, selected and graded in accordance with that Code. Quarter sawn or flat sawn laminations may be used and a mixture of hardwoods and softwoods can be employed by agreement with the purchaser. In practice, the European softwoods are generally employed but stronger laminates at the extremes of the section produce a great structural benefit.

Stress-graded timber is used for its reliability and the outer laminates should be 'selected clear' grades for aesthetic appearance and strength. The size, as already mentioned, varies from 50 mm nominal for straight members down to 20 mm or less for bent members depending on the radius of curvature, bearing in mind that the lamina must be bent in a dry state, unheated and with glue applied. The faces are carefully machined to accurate dimensions to achieve a full contact along the glue lines and the moisture content is controlled to between 8 and 18 per cent depending on the type of adhesive to be used and the final destination of the member.

10.2.2 Adhesives

One of the basic requirements of laminated timber is that the bonding material must be as strong and durable as the timber itself. Timber structures in the past always suffered from weakness at the joints, but with the developments in modern glues, joints now break in the timber rather than along the glue joint when tested to destruction.

As mentioned, the BS 4169 recommends the use of casein or synthetic resin adhesives. Generally, casein glues are easy to mix and apply, and make extremely strong joints, but being derived from sour milk they lose much of their strength when wetted, and are liable to attack by micro-organisms. They should not be used if the timber in use is likely to attain a moisture content over 20 per cent.

The synthetic resin adhesives are plastic with urea, resorcinal or phenolic bases, manufactured in a liquid or powder form or as a dry impregnated film and used either hot or cold. They are immune from attack by moulds and bacteria and are resistant to moisture. There are two principal types: the urea-formaldehyde for interior work and the phenol or resorcinal-formaldehyde for exterior work (see Chapter 20).

Glues used for building structures should be gap-filling; that is to say, they must be capable of bridging the gaps in joints of up to 1.25 mm which can occur due to minor inaccuracies or difficulties over the application of an even pressure. The gap-filling property is obtained by the addition of suitable fillers.

10.3   Laminated beams

The use of a laminated construction for rectangular beams has a number of advantages over a corresponding solid timber section although, because of the labour involved, the laminated member is likely to be more expensive. First, timber being a natural material is subject to variations in strength due to the occurrence of defects, even within a stress graded quality. Any such defects – splits or knots – in a solid member will affect the whole section, whereas the same defect in a laminated member is restricted in its effect to the laminate in which it occurs. Thus the characteristics of the section are much more constant when made up from a lot of small pieces.

Second, seasoning of the thin laminate is easier and more effective as regards strength than for a large section timber and the centre of a laminated beam is virtually as dry as the outer faces. Third, timber boards up to 50 mm thick are readily available, whereas a solid section of a size comparable to that feasible by lamination would be difficult to find. Not only are the small sizes easier to obtain but a laminated beam can use pieces of carcassing timber of a size which otherwise would not find a structural application.

Finally, a laminated beam is not limited in its length by the length of available timber. Laminates can be joined end to end and variations in the cross-sectional area to accord with changes in stress are a feasible proposition in a built-up member, with consequential economies in material.

Further variations based on stress can be introduced by the selection and mix of timbers for the laminates. Any beam under load is subjected to a maximum compressive stress at the top and a maximum tensile stress at the bottom whilst at the point mid-way between these two, the neutral axis, there is no stress at all. Steel beams are 'I' shaped to take account of this stress distribution by concentrating their metal at the top and bottom extremities. Laminated beams can be designed with this in mind by using a strong timber for the upper and lower laminates and a lower grade (and therefore cheaper) timber used for the middle laminates.

The production of a straight laminated beam is relatively straightforward (see Fig. 10.1). The laminates are prepared approx. 10 mm over size in their width but accurately machined in thickness, glued either by brush or by a mechanical glue spreader, assembled as required and pressed together. The pressure required is normally 690 to 1380 kN/m^2 for softwoods and this can be achieved by cramps spaced between 100 and 400 mm apart and tightened down with powered nut runners or a torque wrench. The pressure is maintained

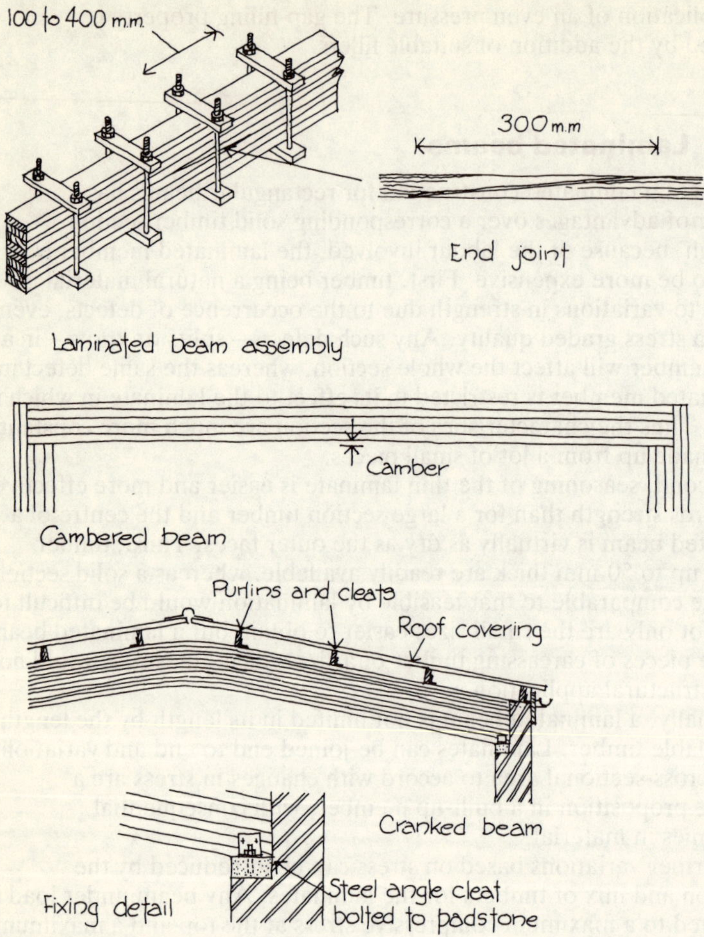

Fig. 10.1 Laminated beams

for six to twelve hours depending on the type of adhesive used, after which the member is planed down to the specified size, sanded, sealed and wrapped in a protective covering.

End joints of laminates is usually by a scarf joint as shown in Fig. 10.1. This drawing also shows a beam cambered to provide greater strength at the mid-span where the maximum bending moment occurs, a cranked beam which produces the desired fall on the roof and details of end fixings by the use of steel cleats and bolts.

10.4 Curved laminated members

It is in the production of curved or bent members that laminated construction really comes into its own. The ease with which attractive shapes can be produced and the high inherent strength they possess appeals both to the designer and the engineer.

The way in which the curved shapes are retained can be seen if two strips of card or plastic of equal length are held together at one end and then bent. One strip slides along the other and their ends become out of line (see Fig. 10.2a). If, in this position, they are then fixed to each other they are prevented from sliding back along each other when the bending force is removed and therefore remain curved. Figure 10.2b shows this principle applied to the production of a curved rib.

There are three principal applications in building work for curved timber laminates: bow-string roof trusses, Portal frames and the bent members of building joinery such as the strings of a helical staircase or the fascia over a bow window or the more complex shapes found in, say, shop-fitting work.

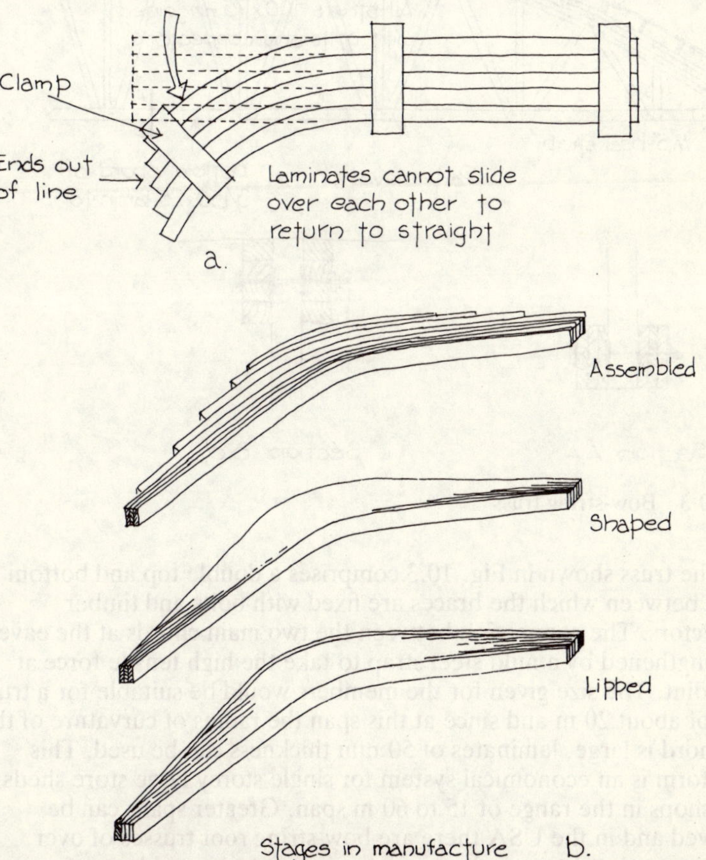

Fig. 10.2 Curved laminated members

10.5 Bow-string truss

Traditionally the bow-string truss as shown in Fig. 10.3 was known as
the Belfast truss but by laminating the curved top member and the
bottom chord a far more efficient structural unit is formed. The shape
of the bowstring truss is such that a very large proportion of the forces
is carried by the curved top member and the bottom chord and the
forces in the braces are small. Consequently these lightly loaded short
members usually are made of solid timber.

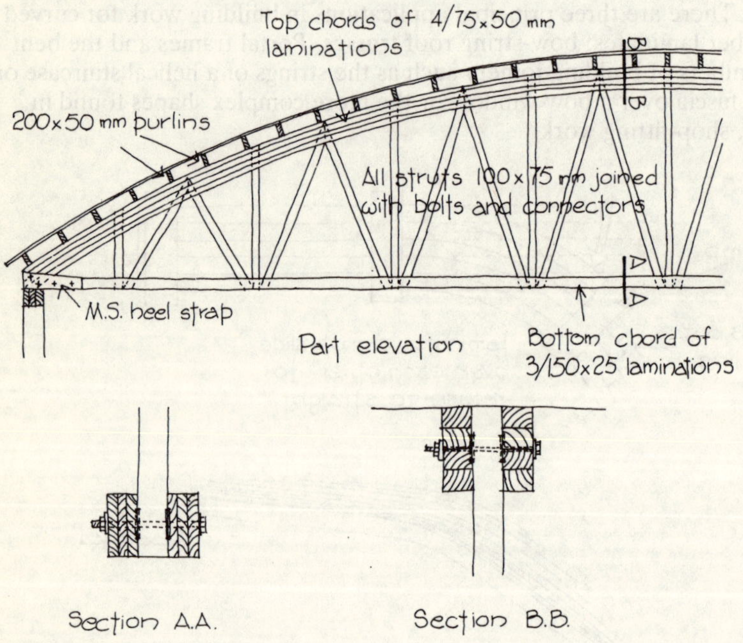

Fig. 10.3 Bow-string truss

The truss shown in Fig. 10.3 comprises a double top and bottom
chord between which the braces are fixed with bolts and timber
connectors. The connection between the two main chords at the eaves
is strengthened by a mild steel strap to take the high tensile force at
this point. The size given for the members would be suitable for a truss
span of about 20 m and since at this span the radius of curvature of the
top chord is large, laminates of 50 mm thickness can be used. This
truss form is an economical system for single storey large store sheds or
workshops in the range of 15 to 60 m span. Greater spans can be
achieved and in the USA there are bow-string roof trusses of over
100 m span, but any very long units present severe problems of
transportation in this country, even in a broken-down form.

10.6 Portal frames

The area of building work in which laminated timber has made the biggest impact is in Portal framed structures, i.e. buildings based on a single bent member which supports both the wall and the roof as shown in Fig. 10.4. The idea is not new: many old cottages dating back to the Anglo-Saxon days employed this principle but used a 'cruck' cut from a tree which had grown at the desired curvature. When the country was covered with dense uncontrolled forests such conveniently

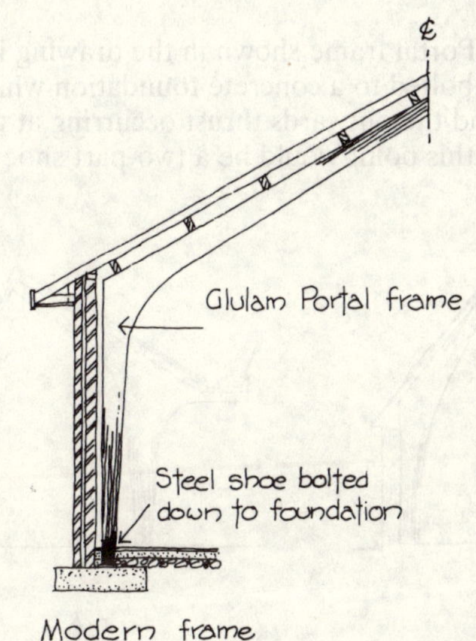

Fig. 10.4 Portal frames

108

bent trees would be readily found, but clearance of the land and careful forestry methods aimed at producing the more marketable straight timber has eliminated this possibility and we must now buy straight timber and bend it.

A great asset of a laminated timber Portal frame is that it possesses an attractive appearance without any further treatment being required other than varnishing. Whilst this means more care is needed in handling and protection after installation it avoids the necessity of subsequent casing or treatment of the frame, as in the case with other materials.

The foot of the Portal frame shown in the drawing is fitted with a metal shoe which is bolted to a concrete foundation which must be designed to withstand the outwards thrust occurring at this point. An alternative detail at this point would be a two-part shoe providing a

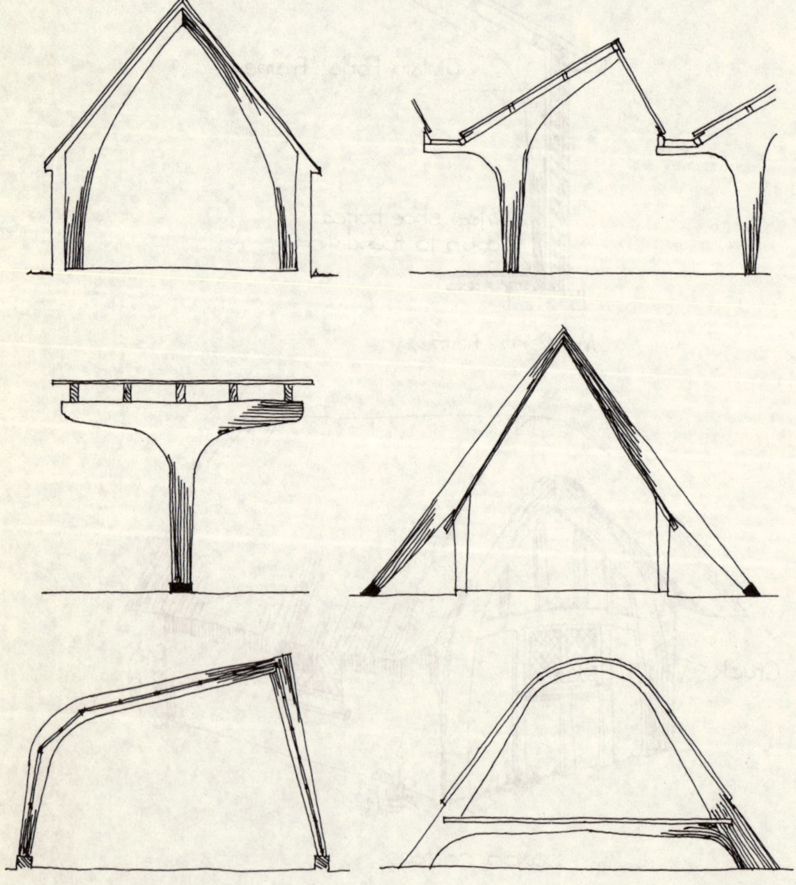

Fig. 10.5   Laminated frame shapes

roller bearing support which would make the structure a three-point pin-jointed frame. This subject is covered more fully in the next chapter.

Connection of the frames at the apex is either by means of a plywood gusset on each face or a handrail type of bolt or alternatively steel shoes and pivots as for the foot of the frame.

The structural shapes which can be achieved by this means are practically limitless. Figure 10.5 shows a few, some of which are not Portal frames but embody similar principles.

10.7 Curved joinery

There are many applications for laminated timber in building joinery, especially where curved shapes are involved, and one such typical application is for the string of a helical staircase as shown in Fig. 10.6. To produce this in solid timber would be very wasteful and, because the line of the grain does not follow the line of the timber, it would be much weaker than its laminated counterpart.

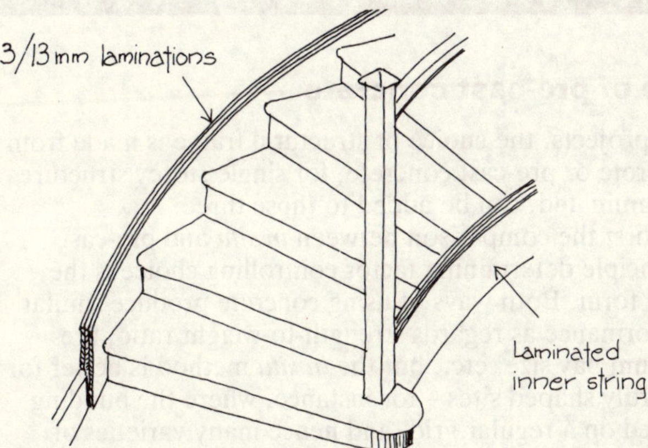

Fig. 10.6 Helical stairs

Chapter 11

Pre-cast concrete frames

11.1 Choice of pre-cast concrete

For multi-storey projects, the choice of structural frame is made from steel, *in situ* concrete or pre-cast concrete; for single storey structures timber, solid or laminated, can be added to those three.

Considering first the comparison between *in situ* and pre-cast concrete, the principle determining factor controlling choice is the building shape or form. Both ways of using concrete produce similar standards of performance as regards strength-to-weight ratio, fire resistance, optimum bay size, etc., but the *in situ* method is better for cramped, awkwardly shaped sites – for instance, where the building cannot be designed on a regular grid, and hence many varieties of beam and column length and size can arise. In this respect, the pre-cast method must be standardised to be economically viable so that it is appropriate for a continuous flow factory production line.

Another advantage of an *in situ* frame is that the columns, beams and floors are one continuous homogeneous structure. The beam to column joint is rigid and therefore each member stiffens the other and the floor acts as a horizontal diaphragm keeping the whole structure rigid, whereas a pre-cast concrete frame must be considered as a pin-jointed structure and the members must achieve their rigidity independently.

Nonetheless, a pre-cast structure has its advantages, the greatest being speed. The moment a beam is secured it can be walked upon (unlike *in situ* work, which must be left to set and cure) and the time taken to install the columns and beams is but a fraction of that required to set up shuttering, fix reinforcement, fill with concrete and then, after due time, strike the shuttering. This on-site speed of erection plus the advantages of continuous production in a controlled environment combine to make pre-cast frames economically attractive for both multi-storey and single-storey work of a simple regular nature.

One controlling factor (and this would apply to steel) is transport and access. The length of the members must be restricted to that which can be transported through the streets to the site and manoeuvred round any sharp corners or other obstructions on the way.

In comparison with steel for multi-storey frames, the pre-cast concrete system is often cheaper because the former has to be provided with a fire-resisting casing which, if it is concrete, can mean as much concrete being used as would be the case if a concrete structure was built, though enclosing more steel than the concrete structure would need for reinforcement.

The advantages of steel over pre-cast concrete are a lower strength/weight ratio which relieves erection problems, and the ease with which another steel member can be attached at a later date should the building require enlargement or modification.

With respect to single storey work, *in situ* concrete is slow and laborious but affords flexibility of design, while steel and pre-cast concrete are swift; but the former may need more finishing which can offset the speed of erection. Laminated timber offers a high aesthetic value but, generally, at a slightly higher cost.

To sum up, a pre-cast concrete framed structure would be likely to be selected for a multi-storey structure where the building is simple in its form, light in its imposed loading and the site readily accessible and with adequate room for delivering, storing and hoisting the pre-cast units. For a single-storey building where an exposed steel frame would not be acceptable but where the standard of finish required is not high (or there is an adequate allowance in the contract to cover the cost of finishing), pre-cast concrete would probably again be used.

11.2 Single-storey framed structure

Fundamentally there is little difference between single- and multi-storey frames insofar as that the structure comprises horizontal members spanning between vertical members, but the fact that the horizontals support a roof rather than a floor and that the verticals do not have to continue up past the horizontals can have a considerable effect on the detail design. For this reason they are being considered separately.

Two aspects of a roof affect the design of the frame: first, the lightness of the imposed load, and consequently the self-weight, as compared with a floor; and second, the necessity to achieve a slope or pitch of the roof surface to direct the rainwater to the desired points.

112

The light load means that smaller sized members can be used. As a result these can be in longer lengths and still remain manageable on site, with consequently fewer joints to be made. The slope or pitch requirement often gives rise to a corresponding pitch in the 'horizontal' members, especially where a flat, level ceiling is not required.

Figure 11.1 shows a typical single-storey building with a flat roof using reinforced concrete columns and beams. This simple form of structure is easy to design, the beam is subject to simple bending, the column loads follow the axis of the section for all practical purposes, the joints between beam and column only need to achieve an adequate bearing and the whole system is statically determinate. Unfortunately, the end result is not necessarily the most economic solution, especially for long spans when the depth required for the beam to resist bending becomes excessive.

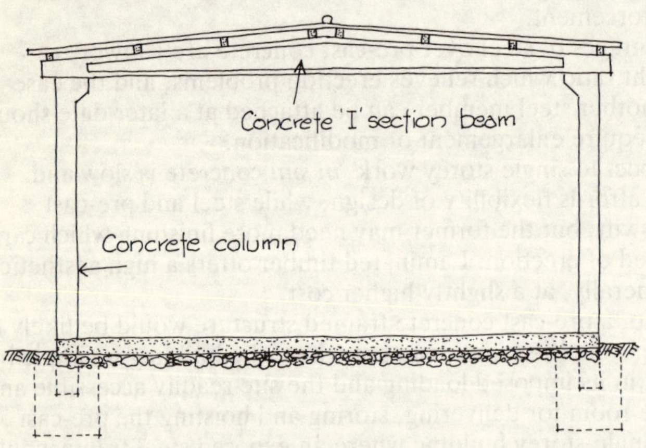

Fig. 11.1   R.C. post and beam building

The simple joint shown in Fig. 11.1 transmits the beam load to the column as a direct downward pressure. It does not induce stresses in any other direction since it is, in effect, a hinge or pin-joint. If a rigid joint is substituted the inherent stiffness of the column relieves the beam of some of the bending stresses because the two act together as shown in Fig. 11.2. A further development of making a rigid joint is to cast the beam or rafter and the column in one piece, producing the Portal frame developed just after the Second World War.

The Portal frame in reinforced concrete (and in steel) is a very popular structural system when large clear spans are required; it combines well with lightweight roofing and cladding sheets, such as corrugated asbestos-cement sheets, to achieve spacious, lofty, low-cost buildings for industrial use. With an upgrading of the finishes the buildings can also be employed as village halls, sports halls, showrooms, etc.

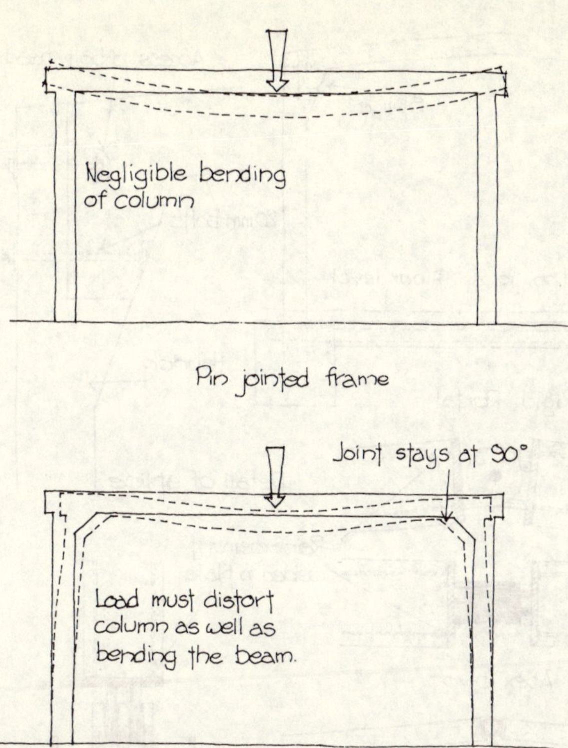

Fig. 11.2 Rigid and pin jointed frames

The design of the joint at the ridge and the detail of the connection between the column and foundation must receive special attention. The frame may be made up in pieces in a variety of ways, either as a fixed frame in which all connections are rigid; as a two-pin frame where the base connections are hinged; or as a three-pin frame where the apex joint is a hinge as well as the base. Figure 11.3 illustrates these forms of frame.

A feature of Portal frames is the unique system of forces at foundation level. Normally foundation pads merely resist a vertical pressure; but in a Portal framed building, the foundations are also subject to a rotational force and an outward thrust. The introduction of the pin joint or hinge between the column and the base prevents this happening but increases the stress in the column. An alternative is either to rely on the passive earth pressure on the sides of the foundation pad or to tie opposite foundations together to resist the outwards thrust; or else to incline the foundation so as to be normal to the line of action of the load and thus all forces are compressive.

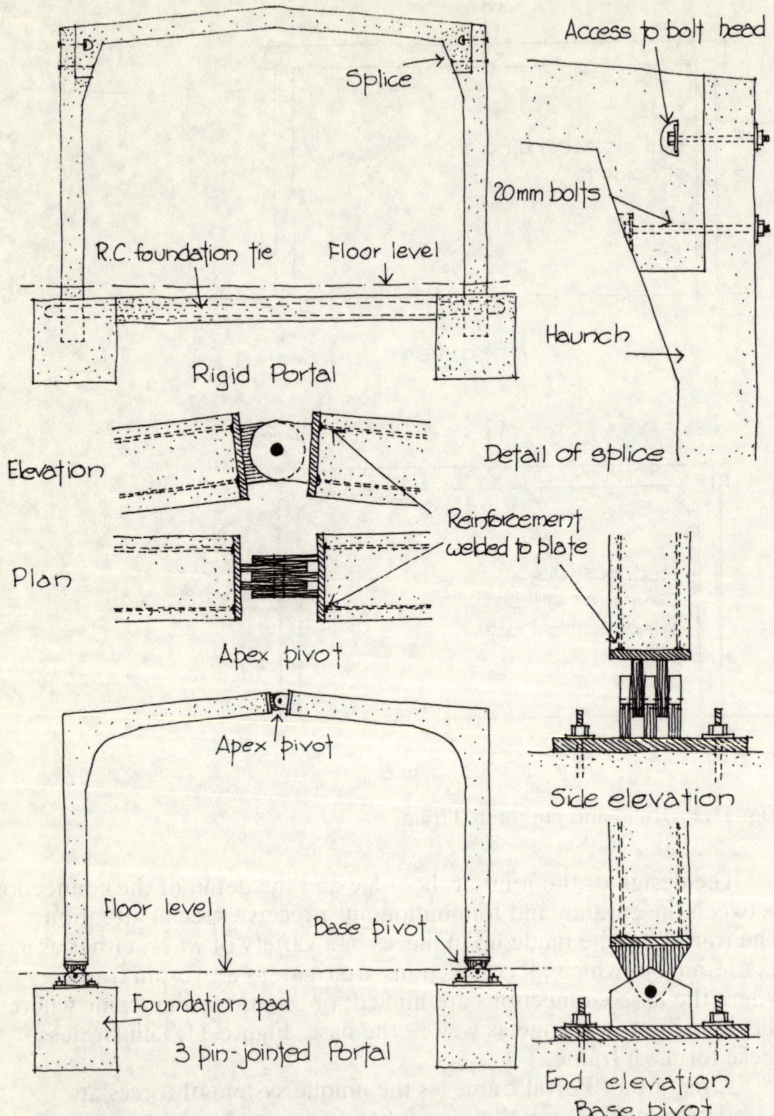

Fig. 11.3 Reinforced concrete Portal frames

11.3 Multi-storey framed structures

A multi-storey pre-cast concrete frame can be assembled on site from a collection of beams and storey height columns. This, however, can give rise to complicated and difficult joints where, say, four beams and an upper and lower column all connect at the same point. Not only is the actual forming of such a joint complicated but each of the six conjoining members must be firmly and accurately held in place until the joint is firm.

To reduce this jointing problem the columns may be pre-cast in lengths of as much as five storeys provided that adequate access and on-site handling is available. With this method, a ledge is provided on which to rest the beams whilst forming the joint and to provide a permanent bearing. This ledge may take the form of a protruding haunch (Fig. 11.4*a*) or, if the presence of the haunch is undesirable in appearance or because it gets in the way of frames or panels fitted to column and beam faces, it may take the form of a housing cast into the column face (Fig. 11.4*b*).

Rather than extending the length of the column to simplify on-site assembly, the uprights can be kept to a one-storey height, but with two

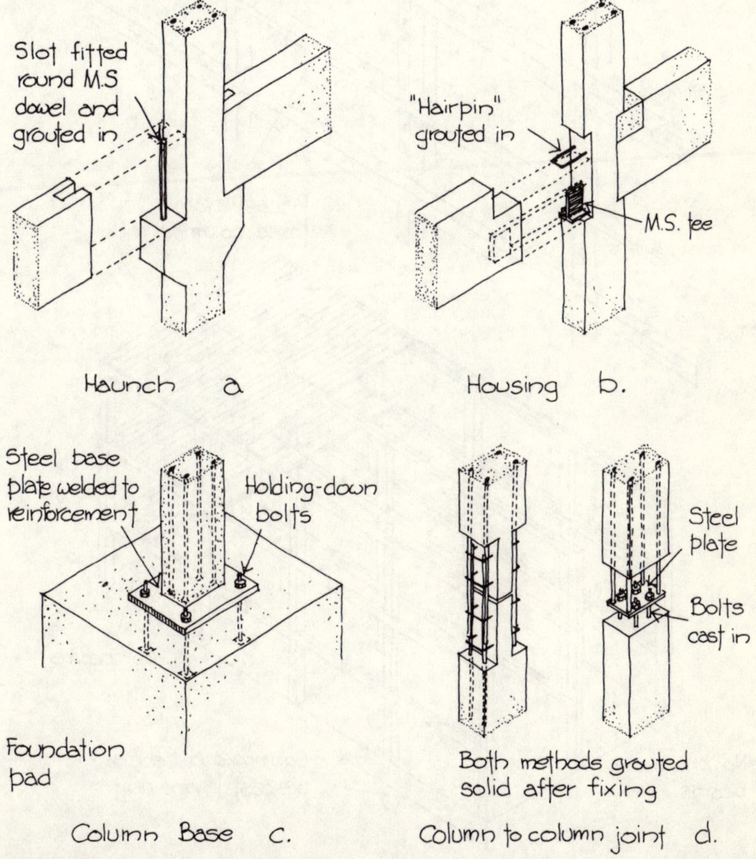

Fig. 11.4 R.C. multi-storey frame joints

or more linked by beams cast in with the columns as a frame unit. This is further improved if the design includes a beam running below window sill level which can also be cast into the unit and will stiffen the columns during handling. The size of such a unit is limited, as for columns, by the ease of access and the facilities for handling, but can be three or four columns wide and/or two or three storeys high. The joint between framed units is arranged to be clear of the cast beam/column connection, as shown in Fig. 11.5.

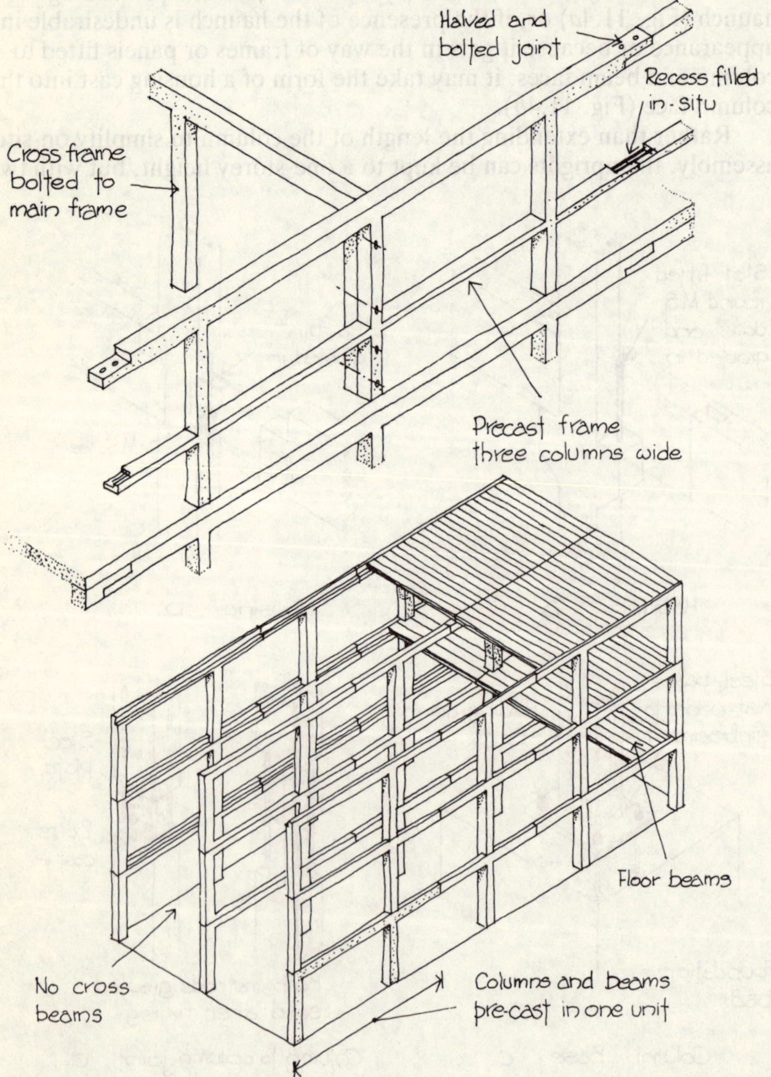

Fig. 11.5 Pre-cast frame joints

For practical reasons of manufacture and transport the beams and columns comprising the unit can only be in one plane. Any cross beams must be separately cast and jointed or made up as part of another frame unit which then produces a composite column (Fig. 11.5). Pre-cast frame units are, therefore, more suitable for structural layouts which require no cross beams, the floor and roof spanning from one line of units to another (Fig. 11.5).

The system to be adopted and the layout to be designed are as varied as the design problems set and the ideas of the engineers solving them, but one interesting variation stems from the principle that the top floor of a multi-storey building is much the same as a single-storey building with the same design conditions applying. With this in mind several engineers have chosen to adopt a single-storey building structure, such as a Portal frame and set it on top of one or more floors of a multi-storey frame as shown in Fig. 11.6.

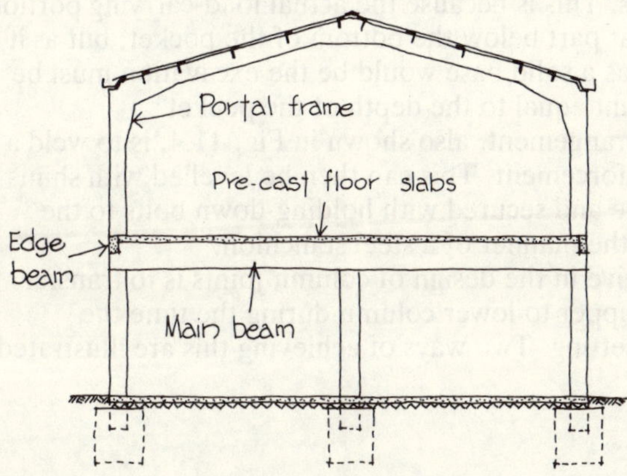

Fig. 11.6 Portal framed top storey

As already mentioned, a drawback to a pre-cast frame is its lack of inherent lateral rigidity. One way of overcoming this weakness is to use diagonal braces fixed in the vertical plan whenever convenient, but as this can cause problems with the plan of the building and create a number of special units it is not a solution often applied.

Two better alternatives exist: using either a solid or hollow pot floor cast *in situ* with the beams so arranged with protruding stirrups that they bond with the floor to produce the continuity which a wholly *in situ* frame enjoys; or else to build parts of the building – a central service core containing stairs, lifts and vertical ducts for instance – as an *in situ* structure and anchor the pre-cast parts of the building to it. The anchoring can be achieved either by bolting the members to steel plates or bolts cast into the core or else by tying the whole structure together with cables passed through specially provided ducts and then tensioned.

118

11.4 Joining pre-cast members

Reference has already been made to Fig. 11.4 showing beam to column connections. The other connections which have to be made are those of the columns to the base and column to column. These are illustrated in Fig. 11.4c and 11.4d.

The base connection is usually effected by forming a pocket in the foundation pad into which the butt end of the column is set and then grouted in. In some cases it is necessary to enlarge the foot of the column to distribute the load from the high grade strong concrete of the frame to the lower grade mass concrete of the foundation.

There are two disadvantages to this construction: first, that it is difficult to obtain an accurately levelled base to the pocket when casting a mass concrete foundation and that levelling shims cannot be inserted once the column has virtually filled the pocket; and second, that the excavation may have to be taken deeper than is necessary for the ground conditions. This is because the actual load-carrying portion of the base is only that part below the bottom of the pocket, but as it must be just as deep as a solid base would be the excavation must be increased by an amount equal to the depth of the pocket.

An alternative arrangement, also shown in Fig. 11.4, is to weld a steel plate to the reinforcement. This can then be levelled with shims placed below the plate and secured with holding-down bolts to the surface of the pad in the manner of a steel stanchion.

The prime objective in the design of column joints is to transfer the load safely from upper to lower column during the time the encasing concrete is setting. Two ways of achieving this are illustrated in Fig. 11.4.

Chapter 12

Infill panels

12.1 Infill panels and cladding

Modern building practice has, in many instances, separated the various functions to be performed by the building fabric and provided elements to satisfy each function separately. In previous chapters one function, that of support, has been examined and a selection of structural elements looked at. Infill panels and cladding represent another group of elements with a different function to perform: that of enclosing space and modifying the external environment to the desired internal conditions.

The difference between infill panels and cladding is that the former fits between the members of the structural frame and thus those members are left exposed, whereas the latter is attached to the face of the frame and conceals it. Infill panels are usually larger elements than cladding because of the necessity to fill the space between the beams and columns. Cladding is usually made up of smaller parts and may comprise pre-cast concrete panels or a framed system of metal mullions and transomes with glass or opaque panels (curtain walling) or it may be one of the many profiled asbestos cement, plastic or metal sheetings attached to sheeting rails fixed to the structural uprights (see Fig. 12.1).

12.2 Functional requirements

In most installations, the infill panels are required to fulfil all the

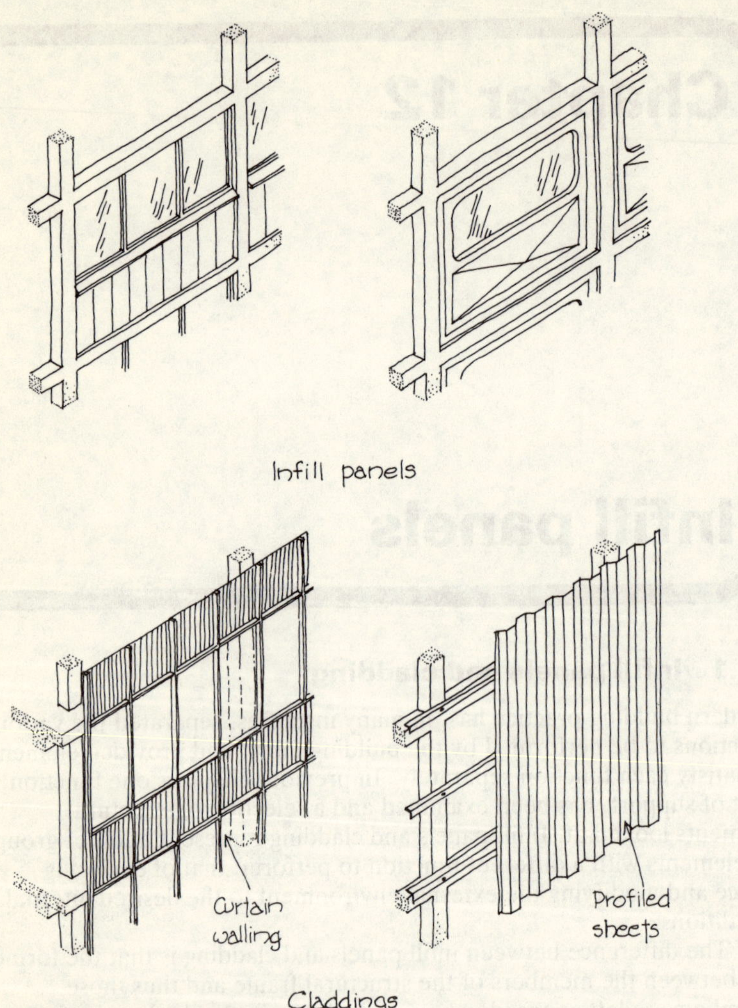

Fig. 12.1 Infill panels and claddings

requirements of the exterior envelope except that of structural support. That is, they must resist the effects of weather, admit the required amount of natural light, be resistant to fire, possess adequate thermal and sound insulating properties, and be draught proof. In these they are sometimes assisted by an inner construction or back-up wall for all or part of their height.

Although an infill panel is not required to afford any structural support to the building it must possess sufficient strength to span between its perimeter fixing points and also to withstand wind pressure. This strength requirement must be allowed for in the framing, in the thickness of the material used as panels and in the method of connecting the panel to the beams and columns.

Being a composite element, an infill panel can be designed to take advantage of the properties of a variety of materials in combination. Thus a material with adequate strength and weather resistance to be used as a panel but low in thermal insulation can be backed up with another material of high insulating properties but which does not possess much strength or weather resistance. In considering the combination of materials and the design of the panel due regard must also be paid to the exposed structural members. Their strength and stability are implied but their appearance, weather and fire resistance, and, particularly, their thermal insulation can be questionable and may need to be enhanced. Typical details dealing with this problem can be seen in the illustrations to this chapter.

12.3　Timber infill panels

Timber is a particularly attractive material for infill panels: it is light but strong, easily worked and joined and can be readily adjusted on site should it be necessary. The panel would, however, suffer from the disadvantages of shrinking, warping, twisting, etc. to which timber is prone and for which allowance must be made in the design. Fire resistance is also a quality lacking in small section timber frames and must be added by the addition of a fire-resisting lining.

Figure 12.2 shows the details of a typical timber-framed panel such as might be used for a block of flats built with a cross-wall structural system. Panels like this were originally made up on site using softwood timber framing, cut and fitted by the carpenter, into which standard window frames would be set and onto which the facing materials would be fixed. The detail shown is similar in principle but intended for off-site prefabrication as a single complete unit or as upper and lower halves if transport or access problems exist.

The construction of the upper, glazed part is similar to an ordinary standard timber window including all the anti-capillary grooves and drips. The lower half demonstrates the way a variety of materials are brought together to contribute their unique properties. First there is the stud frame, made from strong yet light timber, to produce the strength to withstand the forces to which the panel will be subjected; second there is an outer facing; third there is an inner lining and finally there is an infilling within the frame.

The outer lining, in this detail, is also timber, in the form of ship-lap boarding, which would need a finish of paint or some other protection. This is backed up by a layer of felt partly to form a second line of defence and also to act as a draught proofer.

The inner lining is in three layers. The main layer is sheets of 12.5 mm plasterboard nailed to the timber studwork, which is finished

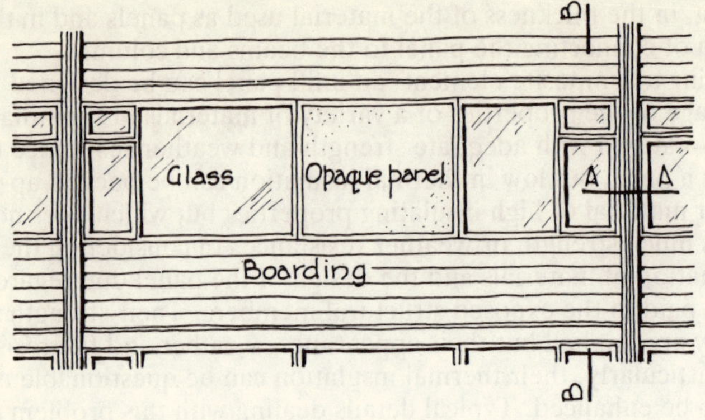

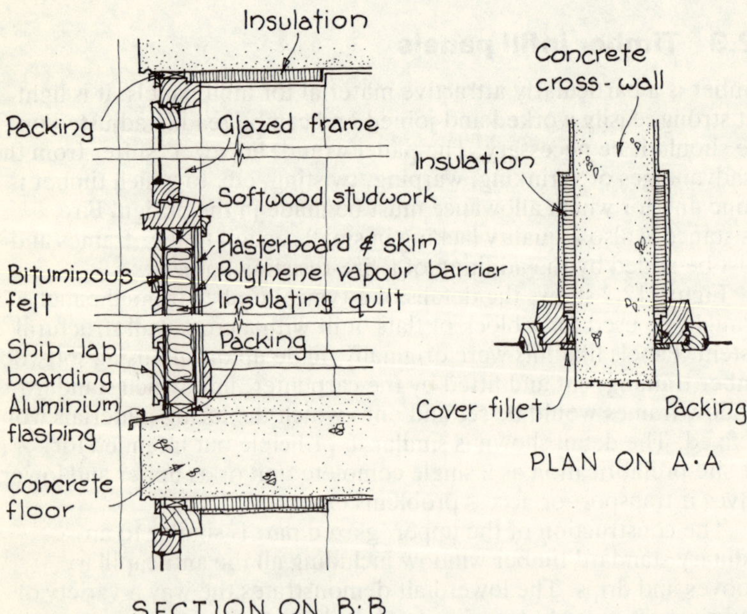

Fig. 12.2 Typical timber frame panel

with a 5 mm layer or skim of gypsum plaster to give the decorative finish. The gypsum skim would probably be site applied to avoid damage and to give a continuous finish with the walls.

These two together contribute the fire resistance of the structure. In this case it would be classed as half-hour fire resisting in accordance with Schedule 8 of the Building Regulations.[1] Between the plasterboard and the studwork is the third layer, the polythene vapour barrier. The positioning of this layer is important; it must always be on the warm side of the insulation since its purpose is to prevent any moisture-laden air from reaching any point in the structure where the temperature is lower than dew-point and condensation can occur. (If the barrier is on the outer side of the insulation, the polythene itself may be cold enough to cause condensation and consequential soaking of the insulation.)

The infilling, in the detail shown, is a thermal insulating quilt of glass fibre or rock wool, necessary to reduce the coefficient of thermal transmission ('U' value) to the required level of either the Building Regulations maximum of 0.6 W/m^2 °C or a lower value if that is desired. With this construction 60 mm thickness of insulation would give a value of about 0.45, 80 mm would give about 0.35 and 100 mm would reduce the U value to 0.3, i.e. half the mandatory standard.

As well as thermal insulation, the infilling may be required to provide sound insulation, in which case either the muffling effect of the thermal insulation may be adequate or, if the noise is from a high energy source, part of the space may need to be filled with a layer of dense sound-absorbing material.

12.4   Metal-framed panels

Figure 12.3 shows a typical detail of an aluminium infill panel. An alternative material to aluminium would be galvanised steel, which is stronger but also heavier and less easy to make up, as the frame is usually assembled first by cutting and welding the sections and then is dipped whole in a hot galvanising bath. Aluminium frames are made up from sections stocked by the supplier which are easily cut with a saw and assembled with screwed junction pieces.

The panel shown in Fig. 12.3 contains all the features described for the timber infill panel – the glazed upper part and the composite lower part – but with a number of variations.

The glazing shown is a sealed double glazing unit to reduce heat loss and condensation, set into the frame with glazing seals and held in place by an inner snap-in glazing bead. This technique of pressing the members together until they snap into position has been developed for aluminium frames where the flexibility of the material can be turned to advantage.

The lower part is finished on the outside with a profiled sheet which may be rolled into ribbed form or pressed into individual raised panel units. The inner finish is of wallboard which can be faced with a great variety of materials, or alternatively many other types of board can be used, depending on the desired appearance. Between the two is a glass fibre insulating mat to reduce heat loss.

The reduction of heat loss via the concrete structure has been achieved in this case by interposing a suitable insulating board between

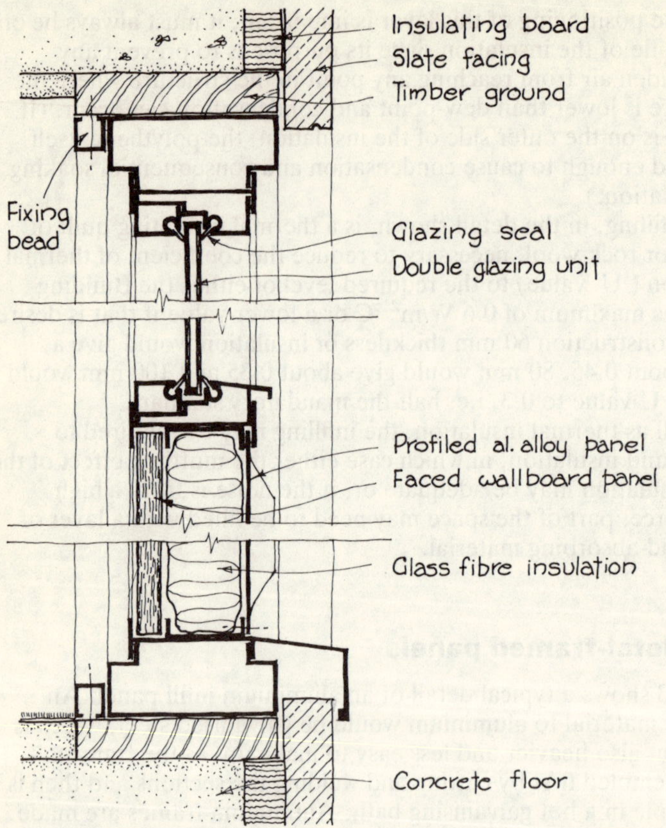

Fig. 12.3 Aluminium infill panel

the concrete and the slate facing. In Fig. 12.2 it is shown applied to the wall face and floor soffite.

One disadvantage of this type of frame is the heat loss through the aluminium which occurs because it is continuous from inside to outside and thus acts as a conductor. Refinements of this simple design to incorporate a discontinuity in the frame and an insulator between inside and outside have now been developed.

To allow the tolerance of fit necessary to cover variations in manufacture of the panel and on-site variations, both in the casting of the concrete and positioning of the panel, the aluminium frame is made smaller than the structural space and the difference is made up with timber fixed to the concrete.

12.5 Cast and moulded panels
The two materials already considered are of a linear or board

formation of predetermined section which are then cut and joined to make up the infill panels. An alternative approach is to use a material which can be cast or moulded and from it construct the main component of the panel. This main component would then have other parts added such as glazing and insulation to complete the panel.

In this category come concrete panels and glass reinforced plastic (GRP) panels. Concrete infill panels are, of course, very heavy and their use is restricted to small size panels (or part panels) and to sites where adequate lifting equipment is already being employed. They can present serious locating and fixing problems but once in place are strong, durable and virtually maintenance free. GRP panels are light and easy to handle and fix, but are not very rigid, are prone to degeneration in sunlight and can present a fire hazard.

12.6 Maintenance

As mentioned above, concrete infill panels should not require any maintenance except, perhaps, cleaning down every 20 years or so and therefore may be selected by the designer for this reason alone, since the maintenance of buildings, particularly the multi-storey structures where infill panels are used, can be greatly affected by the design decisions in the first instance.

In this respect the first of the infill panels shown in Fig. 12.2 is not very good. Being in timber it must receive some form of protective treatment – paint, varnish, etc. – and this often requires renewal at frequent intervals of say every three years. If this sort of infill panel was used on, say, a twenty-storey building the result would be a triennial maintenance nightmare.

For such a large building an aluminium panel could probably present the best choice since the material can be treated to give it a durable finish and the panels are easier to handle than a concrete panel would be.

Note

1 It is proposed to base the 1983 Approved Document AD.A.01 on Regulation E.1 and Schedule 8. (See page 14 above)

Chapter 13

Composite structures

13.1 Definition

Any element or component in a building which makes use of two or
more materials acting together is a composite. The purpose of
combining materials is to make use of the qualities of one to enhance
the performance of the other so that the resulting composite possesses
greater strength than the sum of the strengths of the constituent
materials.

The principle is not new, as can be seen in many old buildings
where the addition of a weak material such as straw or hair to lime
plaster, which is also quite weak, produced a wall coating of
considerable strength and durability. Even the Egyptians knew that
straw in their sun-baked bricks would improve them.

13.2 Examples

There are two ways of using the principle of combining materials for
their greater effect: by placing one within the other, i.e. reinforcement,
or by applying one to the other, i.e. stressed skin construction.

Many examples of reinforcement exist, besides the obvious one of
reinforced concrete. The hair in lime plaster and straw in bricks
mentioned above are both reinforcement in that both the hair and the
straw possess greater tensile strength than the material which contains
them and thus make up for this deficiency. Modern technology is

making use of the remarkable strength of glass fibres in a similar way, by embedding them in a suitable matrix such as concrete or resin, to produce very useful composites and a whole range of new products.

The stressed skin technique is probably less well recognised in building, but anybody who has built a model glider or 'plane using balsa wood and tissue paper has produced one of these composites. The very light but weak balsa wood frame is immensely strengthened by a skin of equally light and weak tissue paper stretched taut across it. Plasterboard is another example: a 9 mm-thick sheet of gypsum plaster would be impossible to handle without breaking and a ceiling lining of just paper would be most unsatisfactory, but if the paper is stuck onto each face of the sheet of plaster the resulting composite is a board of considerable strength and very wide applications.

Considerable reduction can be made in the size of timber floor joists if the decking and the ceiling are of particle board which is glued and screwed to the top and bottom edges of the joists thus stiffening these members and increasing the strength of the complete structure.

13.3 Compatability

For the composite member to remain stable the constituent materials must be fully compatible under all the conditions to which the member might be subjected. There are two principal situations in which the materials must survive together. First, if the two materials have a dis-similar coefficient of thermal expansion, changes in temperature will result in distortion. This is put to use in the bi-metallic strip where two different metals are clamped together and the distortion of the strip due to their different expansion rates used to detect temperature changes. Second, if one material is chemically aggressive to the other in a composite there will be an inevitable failure of the structure. Typical of this is the reaction which occurs between the acids in oak and any steel fixings or components in or on it leading to severe corrosion of the metal and discoloration of the wood.

13.4 Timber and metal composites

A combination of timber and steel or wrought iron used as a structural member is not very common now but in the past the two were put together with great effect.

One method used was to sandwich a wrought iron plate between two timber beams (see Fig. 13.1a) and bolt all three together. By this means the wrought iron strengthened the timber and the timber beams prevented the wrought iron plate from buckling, giving a composite which was far stronger than a solid timber beam of the same overall size. This construction was known as a flitched or flitch-plate beam.

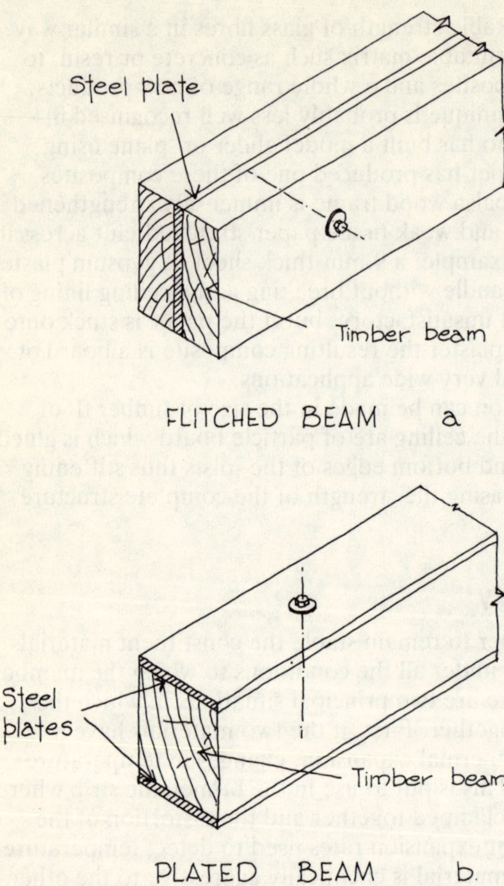

Fig. 13.1 Timber and steel composite beams

In recognition of the fact that the most severely stressed parts of a beam are the top and bottom extreme fibres, another method makes use of steel or iron plates fixed to the top and bottom faces. A plated beam as shown in Fig. 13.1*b* is a good example of a composite since it is not just a strengthened timber beam. It can also be considered that the metal plates carry all the tensile and compressive loads and the timber is there, not as a beam, but as a connection between the two plates making them work together.

13.5 Concrete and steel composites

The most common concrete and steel composite is normal reinforced concrete. It is so widely used now that it is often forgotten that it is a composite and complies with the requirements of a composite in that the resultant member is stronger than the sum of the strengths of the components and these components are compatible.

The strength aspect is certainly true, since the thin steel rods provide the tensile strength that is conspicuously lacking in concrete, so lacking in fact that, except in pre-stressed concrete, the inherent tensile value of the concrete is ignored completely. The steel may also be used to enhance the compressive value of the concrete, and in this case the concrete prevents the thin steel member from buckling, thus allowing it to work at its maximum value.

The chemical compatability of steel and concrete presents no problem and, fortunately, their coefficients of thermal expansion are very near. Concrete expands between 6 and 13 mm $\times$ 10^{-6} per mm length for each °C change in temperature, averaging at 10×10^{-6}, whereas the corresponding figure for steel is 12×10^{-6}. Thus if a reinforced concrete beam 6 m long is subjected to a 40 K rise in temperature the concrete expands 2.4 mm and the steel expands 2.8 mm.

It would appear that the Romans, who used a lot of concrete, attempted to reinforce it but without success, possibly because of this problem of compatability. The metal they had at their disposal was bronze, which has a rate of expansion of around 20×10^{-6} and so, in the beam described above, would expand 4.8 mm – exactly twice the amount of the concrete, which inevitably would disrupt the bond between the two materials leading to structural failure.

The reinforced concrete composite, where the steel is added to the concrete to improve its performance, is not the only form of concrete and steel composite. Many steel-framed buildings have their members encased in concrete, and other structures use a combination of steel and reinforced concrete members in the structural frame.

Until a few years ago the stresses in a steel frame building were calculated, allowing for a concrete casing and a suitable steel section selected, capable of carrying the stress. British Standard 449 now recognises the value of the concrete and steel composite and permits allowances to be made for the structural effect of the concrete casing. The original purpose of the casing was to provide fire resistance but it is now realised that the concrete on the beams stiffens the beam and thus a higher working stress in the steel can be allowed, and on the stanchions it can be allowed to take some of the load thus reducing the size of the steel member (see Fig. 13.2).

Whether the structural frame of a building is steel or concrete, the floors are usually of concrete, generally using a pre-cast system of floor beams resting on the top face of the beams. If the beam is a steel member there is not usually any direct connection between it and the floor, but if a link is made, so that the two act as a composite, the concrete will stiffen the steel and act with it as a beam (see Fig. 13.3). One method of affecting this connection is by shear studs welded onto

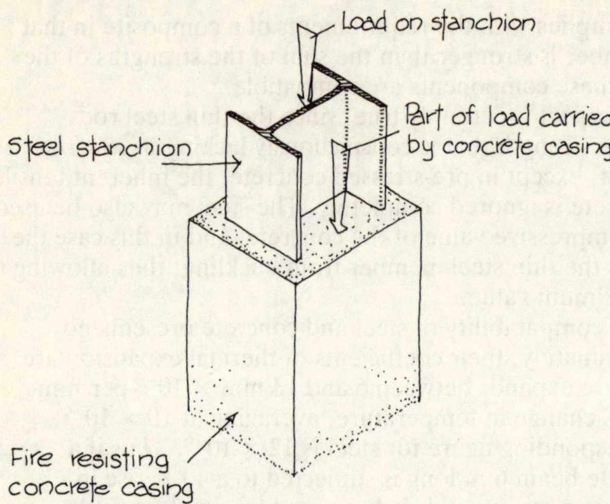

Fig. 13.2 Cased steel member

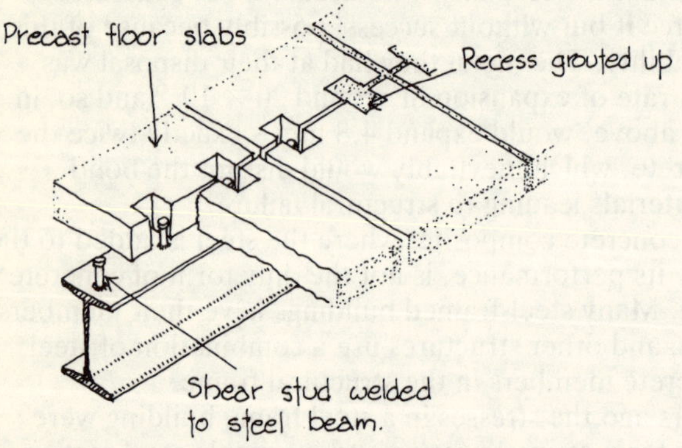

Fig. 13.3 Steel and concrete composite

the top flange of the steel beam, as shown in the diagram. These bond
the materials together and carry the horizontal shear thus creating a
composite member. This technique is covered in BS CP 117 : 1965:
Composite Construction in Structural Steel and Concrete.

13.6 Reinforced brickwork

By embedding steel wires or a strip of expanded metal in the bed joints
of a wall, the brickwork can be given quite remarkable properties. It
can act, for instance, as a composite beam capable of spanning 3 m
with ease and it is also possible to use much larger wall panels within a
structural frame without increasing the thickness. Steel rods built into a

brick pier will add greatly to the pier's stiffness and load-carrying capacity, elevating it to the status of a column.

The procedure to reinforce a wall to act as a beam over an opening is to set up a temporary formwork to support the first course of bricks, lay the first course, spread the bed joint mortar for the next course, press the reinforcement into the mortar and then lay the next course of bricks as normal (see Fig. 13.4).

The design of such a beam follows exactly the same principles as the design of a reinforced concrete beam, but with different values for the safe working loads, and there must be a sufficient number of unbroken courses of brickwork above the opening to provide the

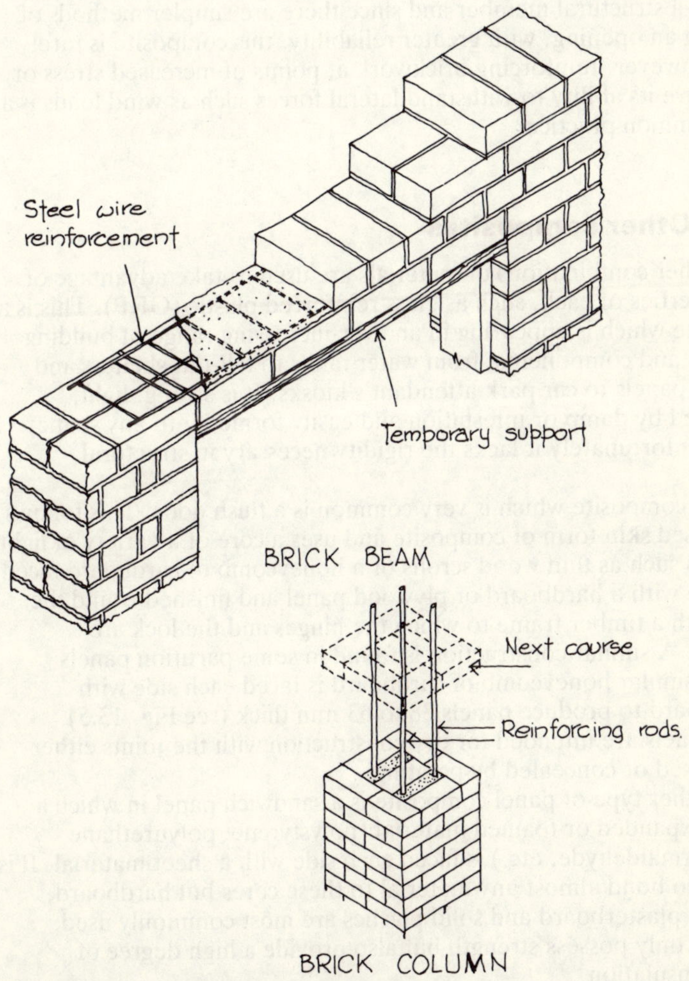

Fig. 13.4 Steel and brick composite

necessary compressive area above the neutral axis of the beam. For large spans more than one course may be reinforced and if there is insufficient cross-sectional area of wall above the opening, compression reinforcement can be placed in the bed joints of the upper courses. Code of Practice 111 : Part 2 : 1970 recommends the design method and values for reinforced masonry.

Reinforced columns are built by carefully arranging the bricks to leave spaces in each course, which are continuous with the one below, and steel rods are inserted into these spaces and grouted in. This main steel reinforcement is connected by stirrups which are bedded in the horizontal joints (see Fig. 13.4).

Much greater care must be exercised in laying bricks if the wall is to act as a structural member and since there are simpler methods of spanning an opening, with greater reliability, this composite is rarely used. However, reinforcing brickwork at points of increased stress or to improve its ability to withstand lateral forces such as wind loads is a fairly common practice.

13.7   Other composites

Many other combinations of materials are used to take advantage of the properties of each, such as glass reinforced plastic (GRP). This is a composite which is appearing in an ever-increasing range of building elements and components, from water tanks to fall-out shelters and cladding panels to car park attendant's kiosks. It is strong, light, unaffected by damp or infestation and easily formed into any shape, though unfortunately it lacks the rigidity necessary in structural elements.

One composite which is very common is a flush door. This follows the stressed skin form of composite and uses a core of a variety of light materials such as thin wood scrolls or a honeycomb of cardboard faced each side with a hardboard or plywood panel and finished round the edges with a timber frame to which the hinges and the lock are attached. A similar construction is found in some partition panels where a similar honeycomb of cardboard is faced each side with plasterboard to produce panels 38 to 63 mm thick (see Fig. 13.5). These panels are intended for dry construction with the joints either left exposed or concealed by painting.

Another type of panel composite is a sandwich panel in which a core of expanded or foamed material (polystyrene, polyurethane, phenolformaldehyde, etc.) is faced each side with a sheet material. It is possible to bond almost any material to these cores but hardboard, plywood, plasterboard and solid plastics are most commonly used. They not only possess strength but also provide a high degree of thermal insulation.

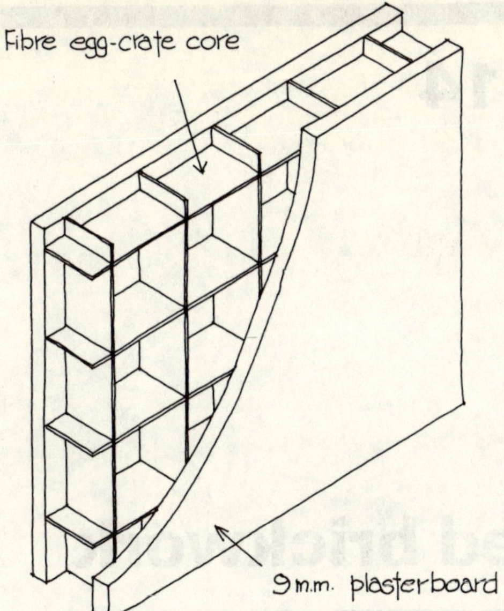

Fig. 13.5 Plasterboard partitioning

Chapter 14

Calculated brickwork

14.1 Design objective

The selection or calculation of the details of a load-bearing brick structure should, and indeed must, comply with the simple statement contained in BR D8, which states: 'The structure of a building above the foundations shall safely sustain and transmit to the foundations the combined dead load, imposed load and wind load without such deflection or deformation as will impair the stability of, or cause damage to, the whole or any part of the building'.[1]

The manner by which the design sets out to achieve this is dependent on many factors. If the project is small and simple, i.e. a residential building of not more than three storeys or a non-residential building not exceeding 9 m span and height, the appropriate wall thickness can be found by reference to Schedule 7 of the Regulations.[2] In any other case (and also in the cases mentioned if unusual conditions are present) the capacity of the wall to meet the Regulation requirement must be calculated in accordance with CP 111 : 1970 : *Structural Recommendations for Load Bearing Walls* (which will eventually be superseded by BS 5628).

14.2 Strength of brickwork

In a wall built of bricks and mortar, the strength depends primarily on the strengths of these two components but will also be influenced by

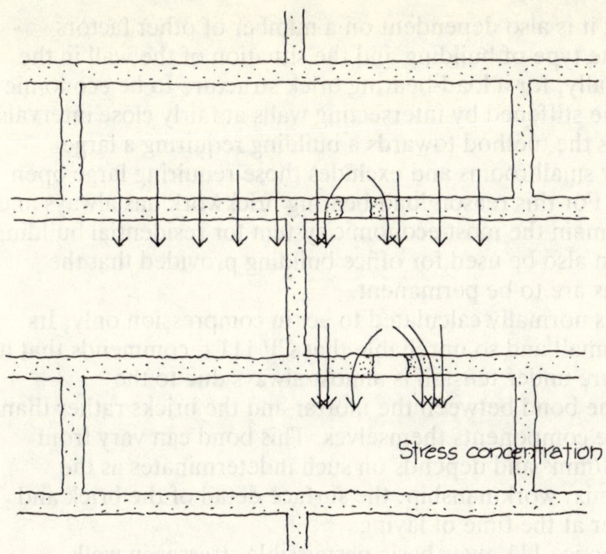

Fig. 14.1 Concentration of stress due to faulty mortar bed

the quality of work. It is clear that unless each brick is supported 100 per cent by the mortar below concentrations of stress will occur (see Fig. 14.1). Furthermore, failure to flush up the perpends, or vertical joints, will impair the distribution of load through the wall.

The strength of the mortar must be matched to the strength of the bricks to try to achieve a uniform load-carrying capacity. The lower limit of load bearing is set by the weaker of the two and whilst the strength of the wall increases as the strength of the brick goes up it is not a directly proportional increase, due to the effect of the mortar.

Brick strength is taken from tests on a random sample of the ten bricks in accordance with BS 3921 and the particular brick graded according to its average crushing strength. Mortar mixes recommended for various grades are shown in Fig. 14.2.

Not only is the strength of a brick wall related to that of its

BRICK TYPE	GRADE N/mm²	CEMENT vol.	LIME vol.	SAND vol.
Low strength	10·35	1	2	9
Medium strength	20·7 to 34·5	1	1	6
High strength	48·3 or more	1	¼	3

Fig. 14.2 Recommended mortar mixes

components but it is also dependent on a number of other factors deriving from the type of building and the situation of the wall in the building. Generally, for a load-bearing brick structure to be economic the walls must be stiffened by intersecting walls at fairly close intervals. This predisposes the method towards a building requiring a large number of fairly small rooms and excludes those requiring large open internal spaces. For this reason load-bearing brickwork has always and will probably remain the most economic system for residential building and flats and can also be used for office building provided that the internal divisions are to be permanent.

Brickwork is normally calculated to act in compression only. Its tensile value is small and so unreliable that CP 111 recommends that it is ignored. Failure under tension is almost always due to the breakdown of the bond between the mortar and the bricks rather than the failure of the components themselves. This bond can vary from 0.28 to 0.55 MN/mm^2 and depends on such indeterminates as the precise mortar mix, workmanship, the surface detail of the brick and even the weather at the time of laying.

Code of Practice 111 gives basic permissible stresses in walls related to the strength of the bricks and type of mortar. The basic stress is assumed to be due to the combined effect of uniformly distributed dead and superimposed loads. These values are shown in the table in Fig. 14.3.

Mortar mix			Brick crushing strength N/mm^2								
Cement	Lime	Sand	2·8	7·0	10·5	20·5	27·5	34·5	52·0	69·0	96·5+
vol.	vol.	vol.				Basic stress N/mm^2					
1	0-¼	3	0·28	0·70	1·05	1·65	2·05	2·50	3·50	4·55	5·85
1	½	4½	0·28	0·70	0·95	1·45	1·70	2·05	2·80	3·60	4·50
1	1	6	0·28	0·70	0·95	1·30	1·60	1·85	2·50	3·10	3·80
1	2	9	0·28	0·55	0·85	1·15	1·45	1·65	2·05	2·50	3·10
1	3	12	0·21	0·49	0·70	0·95	1·15	1·40	1·70	2·05	2·40
	1*	2	0·21	0·49	0·70	0·95	1·15	1·40	1·70	2·05	2·40
	1#	3	0·21	0·42	0·55	0·70	0·75	0·85	1·05	1·15	1·40

* Hydraulic lime
\# Non-hydraulic lime

Fig. 14.3 Basic stress factors

14.3 Loading of walls

The majority of the load on a wall is transferred to it by the floors and the roof. These derive their load from their own mass and from the loads placed upon them. In the case of floors these imposed loads are due to furniture and equipment and also to the occupants, and in the case of a roof due to snow. The mass of the floor and roof does not change and therefore is referred to as 'dead' load, and the actual value can be calculated and used.

The imposed loads are indeterminate both in their position or magnitude and can either change at intervals or, as in the case of the occupants, can move around all the time. To ensure sufficient strength in both the floor and the wall nominal values are assumed, based on the type of occupancy, and applied to the whole floor area. These values are set out in CP 3 : Chapter V : Part 1 : 1967.

In addition to the uniformly distributed loads described above, the wall may be subjected to loads concentrated at one point as a result of beams bearing on the brickwork. Code of Practice 111 lays down that where a point load occurs combined with a uniformly distributed load the local stress must not exceed the basic stress as shown in Fig. 14.3 by more than 50 per cent. This increase can be allowed because the basic stresses are derived from the crushing strength of a brick rather than of brickwork and each individual brick is strengthened by the others, added to which the load is dispersed through the wall below the point of local stress concentration.

As already mentioned, CP 3 : Chapter V : Part 1 sets out live loads to be calculated on each unit area of floor but it is recognised that in a multi-storey building it is not likely that every floor will be loaded to the maximum and therefore it is uneconomic to allow for this total load on the walls. The Code therefore sets out percentage reductions of the total imposed load related to the number of stories in the building. This is shown in Fig. 14.4.

Reference has already been made to the inability of brickwork to withstand tensile forces and it could be thought that this is no problem since these forces are unlikely to occur. Such is not the case because of the way that most walls are loaded eccentrically.

Number of floors carried	Reduction in imposed load %
1	0
2	10
3	20
4	30
5 to 10	40
over 10	50

Fig. 14.4 Reduction of floor loads

138

If a floor passes over a wall and has an equal span on both sides of the wall then the line of action of the load is directly down the axis of the wall (Fig. 14.5a). However, if the floor stops at the wall the line of action of the load is away from the axis of the wall and this eccentricity tends to cause the wall to bend (Fig. 14.5b). Even if the floor bearing is carried right across the wall, flexibility of the floor will cause an eccentricity of up to one-sixth of the bearing width. At this distance the

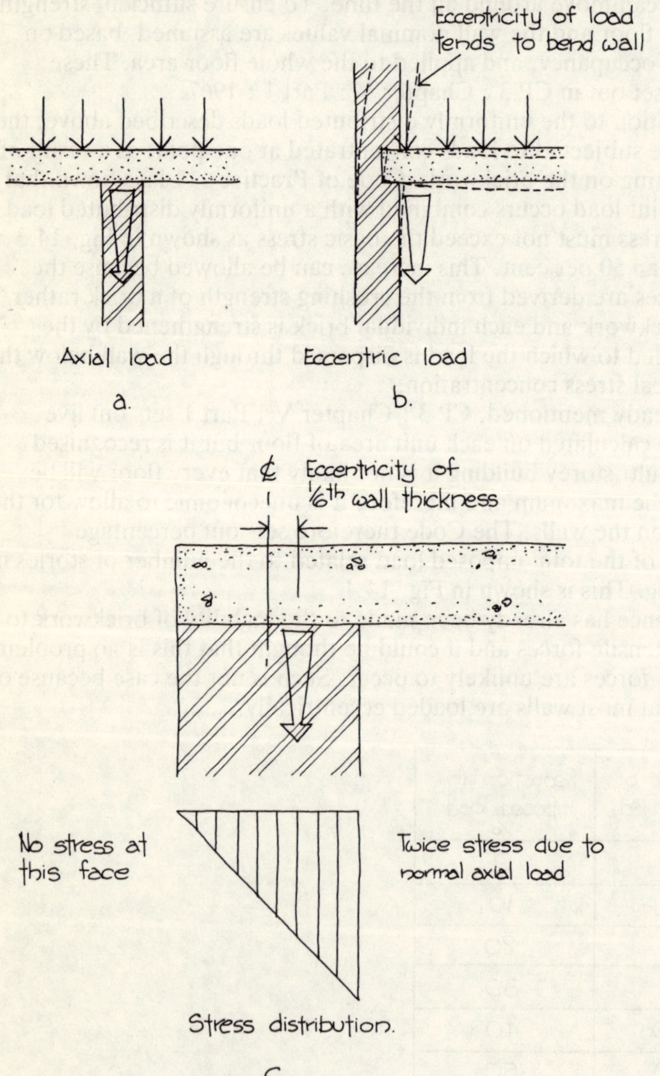

Fig. 14.5 Eccentric loading

change in intensity of the stress is such that the face of the wall opposite to the eccentricity is not subjected to any stress and the other face is required to carry twice the stress which would occur if the load was axially disposed (Fig. 14.5*c*).

Although the primary loads on a wall are due to dead and live loads, there is a third load to be allowed for, the wind load. Assessment of the wind load is by the methods given in CP 3 : Chapter V : Part 2 : 1972 which uses a 'basic wind speed', adjusts it to take account of features such as the height and shape of the building and surrounding topography and then converts it to 'dynamic pressure'. This is used to find the actual pressure or suction on the building surfaces.

It is only the walls on the windward side which are subjected to an inward pressure, those on the leeward side and at the ends of the building receiving an outward suction. These tends to bend the wall inwards or outwards and set up corresponding compressive and tensile forces. The tensile forces may be counterbalanced by the forces due to the imposed loads but the worst combination where the compressive force and loading force coincide must be used in calculation.

14.4   Terms and definitions

In calculating the value of brickwork certain values must be ascertained and employed as follows:

14.4.1   Slenderness ratio

Failure due to bending will always occur in the direction of the least horizontal dimension and the greater the height in relation to this dimension the more easily will the wall or column bend. The slenderness ratio expresses the relationship between the height and thickness and is actually the ratio between the effective height and the effective thickness, both of which are defined below.

Obviously, because of this slenderness a high thin wall or column cannot be expected to carry as much load as a low thick wall or column built from the same bricks. This is reflected by a reduction factor by which the basic stress is multiplied resulting in an increase in the wall thickness. Where the slenderness ratio is less than 6 the full basic stress can be used, but above this the reduction factors as shown in Fig. 14.6 are applied. Figure 14.6 also gives the maximum ratios laid down in CP 111. Where the effective width of a wall is less than its effective height the slenderness ratio is based on this width rather than the height.

14.4.2   Effective height

In many cases, the effective height, i.e. the height used in calculations, is not the actual height but an adjusted value based on the degree of

Type of wall	Max. slenderness ratio	Slenderness ratio	Stress reduction factor			
			Axially loaded	Eccentricity of vertical loading as a proportion of the thickness of the member		
				⅙	¼	⅓
Brickwork in hydraulic lime mortar →	13	6	1·00	1·00	1·00	1·00
		8	0·95	0·93	0·92	0·91
		10	0·89	0·85	0·83	0·81
		12	0·84	0·78	0·75	0·72
		14	0·78	0·70	0·66	0·62
Ditto under 2 storeys and any wall under 90 mm thick over 2 storeys →	20	16	0·73	0·63	0·58	0·53
		18	0·67	0·55	0·49	0·43
		20	0·62	0·48	0·41	0·34
		22	0·56	0·40	0·32	0·24
		24	0·51	0·33	0·24	–
		26	0·45	0·25	–	–
Maximum for any wall →	27	27	0·43	0·22	–	–

Fig. 14.6   Slenderness ratio reduction factors

lateral support or restraint the wall or column enjoys. Where the degree of support is high as with floors built into a wall, the effective height is taken as three quarters of the actual height; whereas at the other extreme the effective height of a brick column with no effective restraint at the top is twice the actual height. These are illustrated in Fig. 14.7.

14.4.3   Effective thickness

This is the actual thickness of a wall, not including finishes such as plaster or rendering, multiplied by a factor based upon the size and spacing of any supporting piers (see Fig. 14.8). With cavity walls the thickness is taken as two thirds of the sum of the nominal thicknesses of the leaves. Where buttressing is by intersecting walls, they are considered as piers of a width equal to the intersecting wall width and a thickness equal to three times the thickness of the buttressed wall (see Fig. 14.8). For a column the effective thickness is taken as the least dimension.

14.4.4   Effective length

The effective length of a wall is taken as the distance between supports

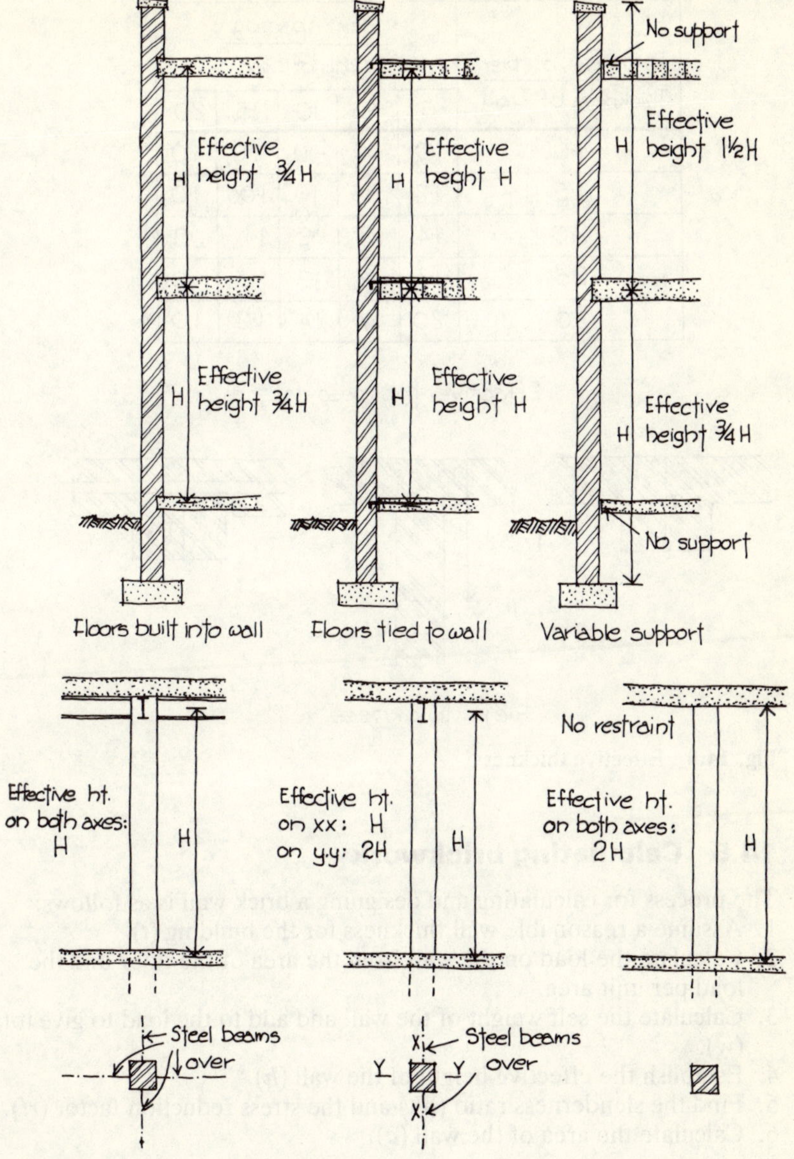

Fig. 14.7 Effective heights

in the form of piers or intersecting or return walls. Where there is no support, as may occur with an end panel of wall, the effective length is two and a half times the actual length from the face of the pier to the end of the wall.

Thickness of pier Thickness of wall	Pier spacing Width of pier				
	6	8	10	15	20
1·0	1·0	1·0	1·0	1·0	1·0
1·5	1·2	1·15	1·1	1·05	1·0
2·0	1·4	1·3	1·2	1·1	1·0
2·5	1·7	1·6	1·3	1·15	1·0
3·0	2·0	1·7	1·4	1·2	1·0

Effective thickness factors

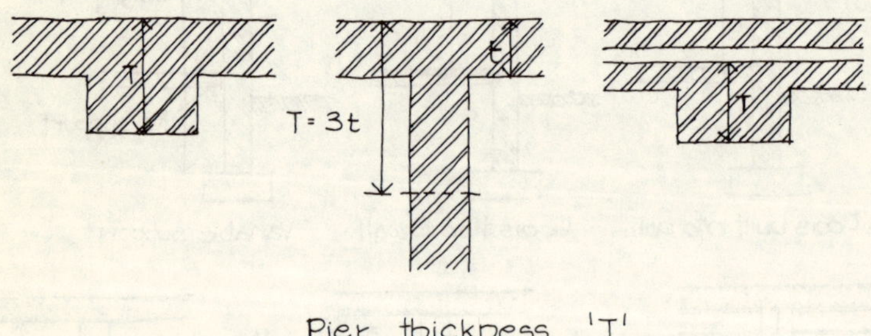

Fig. 14.8 Effective thickness

14.5 Calculating brickwork

The process for calculating and designing a brick wall is as follows:
1. Assume a reasonable wall thickness for the building (t).
2. Calculate the load on the wall from the area of the floor and the load per unit area.
3. Calculate the self weight of the wall and add to the load to give total (w).
4. Establish the effective height of the wall (h).
5. Find the slenderness ratio (h/t) and the stress reduction factor (rf).
6. Calculate the area of the wall (a).
7. Calculate the equivalent basic stress from the formula $\dfrac{w}{a \times rf}$.
8. Select a grade of brick and mortar to suit the equivalent basic stress – Fig. 14.3.

Example

An interior cross wall in cavity work for a five-storey block of flats: the height from floor to floor is 2600 mm, the floors are of precast concrete beams and run right through the wall at each level. The total load on the wall including self weight of brickwork (w) is 140 000 N/m.

Effective height of wall (Fig. 14.7)(h) = 0.75 × actual height
h = 2600 × 0.75
  = 1950 mm.

Thickness of wall (t) = 2/3 × sum of thickness of leaves
  = 2/3 × (102 + 102)
  = 136 mm.

$$\text{Slenderness ratio} = \frac{h}{t}$$

$$= \frac{1950}{136} = 14.35.$$

Stress reduction factor (Fig. 14.6) rf = 0.771 (interpolated).
Area of brickwork (a) = (102 + 102) × 1000 mm^2 per metre run of wall
  = 204,000 mm^2.

$$\text{Equivalent basic stress} = \frac{w}{a \times rf}$$

$$= \frac{140\ 000}{204\ 000 \times 0.771}$$

$$= 0.89 \text{ N/mm}^2$$

From Fig. 14.3:
Bricks with a crushing strength of 10.5 N/mm^2 in a 1 : 1 : 6 cement/lime/sand mortar give a basic stress of 0.95 N/mm^2 which would be suitable or if a stronger brick has been chosen such as one with a crushing strength of 20.5 N/mm^2 a mortar mix of 1 part cement to three parts lime to twelve parts sand would produce the required strength.

Notes

1. It is proposed to transfer this statement, with very little alteration, to B.02 of the 1983 Regulations. (See page 14 above)
2. It is proposed to embody the Rules of Schedule 7 in the 1983 Approved Document AD.B.01. (See page 14 above).

Chapter 15

Long span roof frames

15.1 Roof systems

The materials from which roof frames are constructed comprise
timber, metals (mainly steel or aluminium), concrete and, to a limited
extent, plastics, and these are employed in a very wide range of roof
framing systems of which the following is a small selection.

It is possible to construct any of the framed systems in most of the
materials used but some roof systems (not dealt with in this chapter),
such as vaults or shells, are best in concrete and others, such as
air-supported roofs, require the unique properties of plastics.

15.2 Trussed roofs

Strictly, any statically determinate assemblage of struts and ties
connected together at their ends for the purpose of supporting a roof
covering can be termed a truss but the word is usually restricted to
triangulated plane frames such as are shown in Fig. 15.1. Even with
this restriction the finished shape of the roof is very variable depending
on the building requirements.

The simple 'shed' frame (Fig. 15.1a) of triangular trusses
supported on columns and connected by purlins is a very common form
of roof for small and medium size buildings, but where the span
becomes large, e.g. over 15 m, the area of roofing required and the
volume within the roof which needs to be heated makes it an

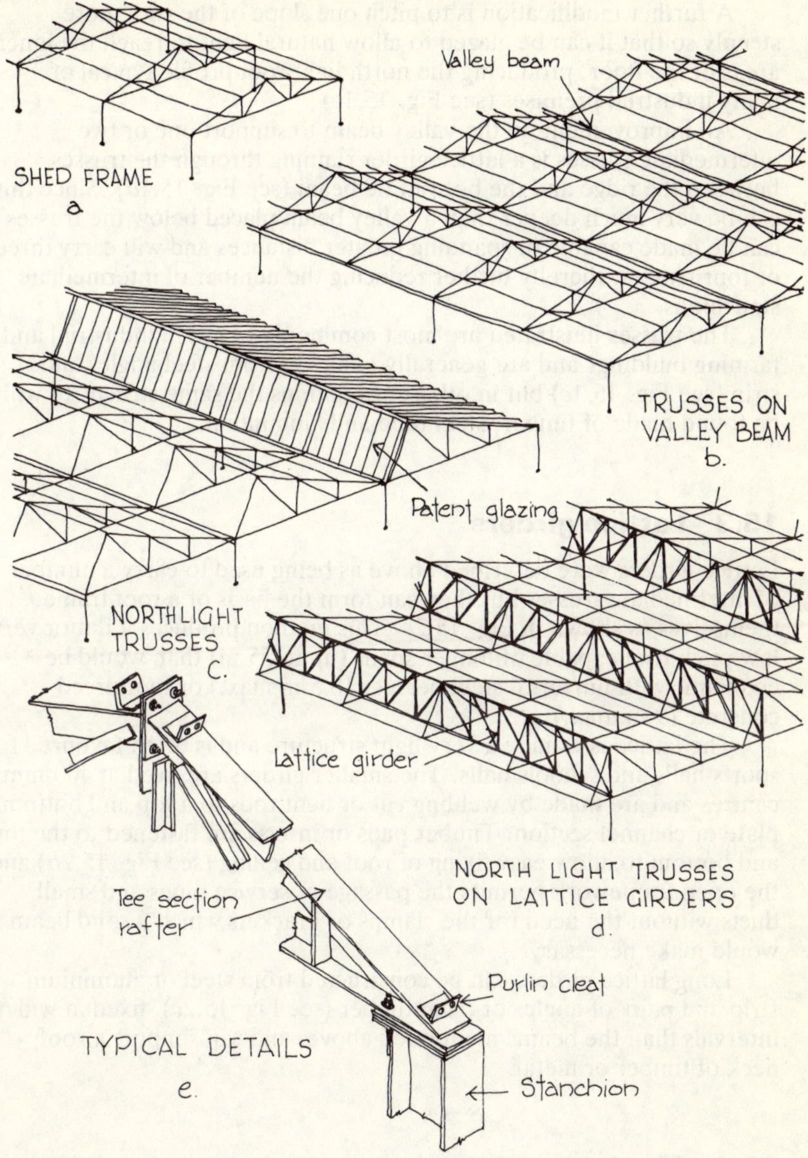

Fig. 15.1 Trussed roofs

uneconomic method. In this case the solution is to use smaller trusses side by side. To reduce the inconvenience of a lot of intermediate columns obstructing the floor space a valley beam can be introduced to carry one or more intermediate trusses (see Fig. 15.1*b*).

A further modification is to pitch one slope of the roof more steeply so that it can be glazed to allow natural light to reach the inner areas of the floor, producing the north light roof profile typical of many industrial premises (see Fig. 15.1*c*).

An improvement on the valley beam to support one or two intermediate trusses is a lattice girder running through the trusses between the ridge and the bottom tie beam (see Fig. 15.1*d*). Since this can be very much deeper than a valley beam placed below the trusses it can be made capable of spanning greater distances and will carry three or four trusses, thereby further reducing the number of intermediate stanchions.

The trusses illustrated are most commonly found in industrial and farming buildings and are generally made up from steel angles and strip (see Fig. 15.1*e*) but in other installations the same structures will be found made of timber, steel tube and aluminium.

15.3 Lattice girders

Lattice girders were described above as being used to carry a number of intermediate trusses but they can form the basis of a roof frame themselves as shown in Fig. 15.2*a*. This method produces a flat or very low pitch roof capable of longer spans (up to 35 m) than would be economic with universal steel beams (10.5 m max.) or reinforced concrete (9 m max.).

The lattice is usually a very light structure and is often favoured for sports halls and schools halls. The smaller girders are fixed at 400 mm centres and are made by welding cut or bent rods to a top and bottom plate or channel section. Timber pads or inserts are fastened to the top and bottom to allow easy fixing of roof and ceiling (see Fig. 15.2*b*) and the open framework permits the passage of service pipes and small ducts without the need for the clamps or brackets which a solid beam would make necessary.

Long lattice girders can be constructed from steel or aluminium strip and pairs of angles or from timber (see Fig. 15.2*c*), fixed at wider intervals than the beams mentioned above, and can support a roof deck of timber or metal.

15.4 Monitor roofs

The monitor roof was developed from the flat roof in the same way and for the same reasons as the north light truss was developed from the symmetrical truss – to permit natural light to reach the centre of the roofed areas. It consists of what is basically a flat roof but it is interrupted at regular intervals by rows of monitor lights which are raised sections of roof with vertical, or near vertical, glazing each side

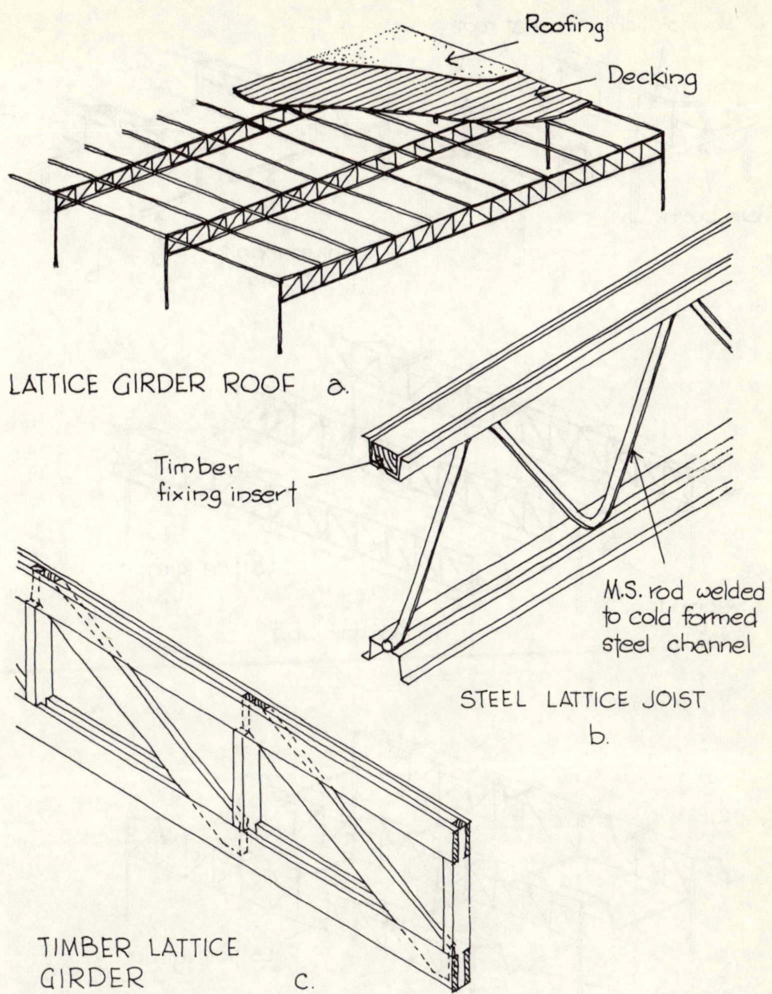

Fig. 15.2 Lattice joist and girder roofs

(see Fig. 15.3). The advantages of this arrangement are that the same effect as a north light truss is achieved but with less roof volume, and the flat roof areas between the monitor lights make access for maintenance a great deal easier. (One of the drawbacks to any system of roof lights is that they tend not to be cleaned as regularly as their position requires and consequently the amount of natural light admitted gradually diminishes.)

There are several methods of construction, generally using steel or aluminium, but pre-cast concrete monitor lights are available to be used in conjunction with similarly formed columns and beams.

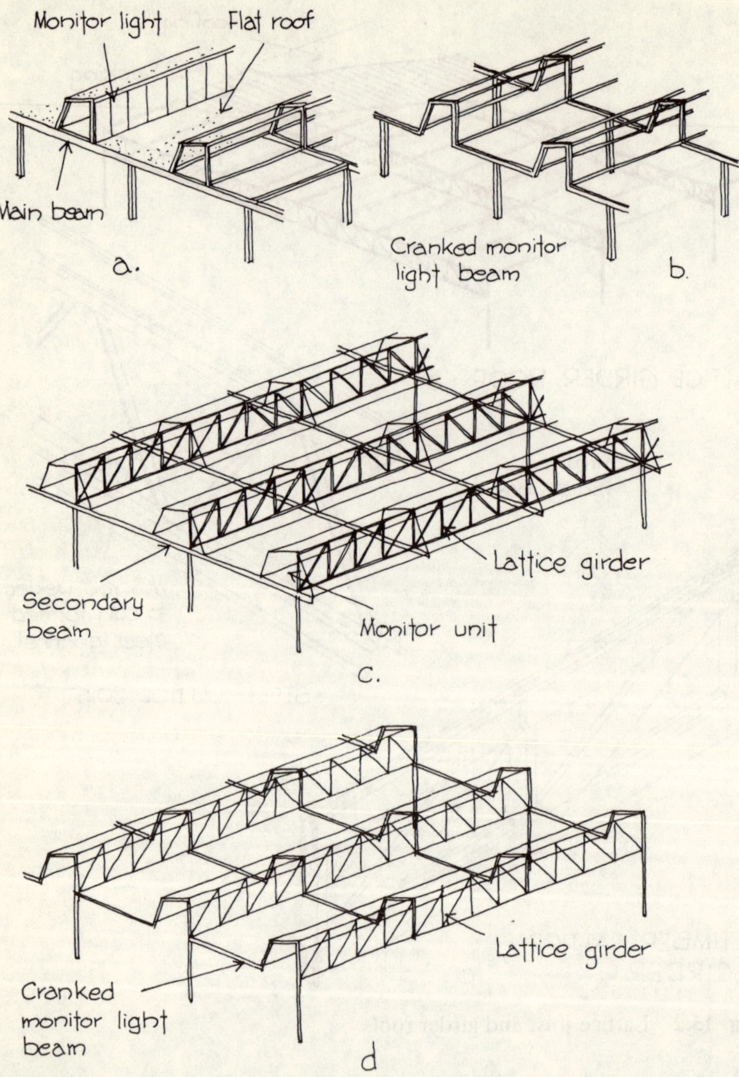

Fig. 15.3 Monitor roof lights

The simplest method is to span across the tops of rows of
stanchions with beams and across these beams place the monitor units,
alternating with panels of flat roofing (see Fig. 15.3*a*). This has an
unattractive feature in the beam which continues across below the
monitor light. To avoid this, cranked beams are used which still span
from stanchion to stanchion but follow the profile of the roof (see
Fig. 15.3*b*).

Using straight or cranked beams means that the stanchions must be spaced at an economic distance for a beam section. This can result in an unacceptable degree of obstruction, so to reduce the number of stanchions, lattice girders can be introduced as for trussed roofs, extending up into the monitor lights. The roof system can then be supported on secondary beams bearing onto the bottom member of the lattice girder (see Fig. 15.3*c*) or by cranked beams following the roof profile and connecting the bottom member of one girder to the top member of the next (see Fig. 15.3*d*).

15.5   Flat grid roofs

If, instead of providing a number of large primary beams across which span a larger number of smaller beams, all the beams are made the same size and spacing and fixed to each other at the same level, a rectangular grid results in which each member supports the other (see Fig. 15.4*a*) producing a structure of great strength. The method can be further improved, firstly by turning the grid to run diagonally so that the members follow the lines of stress more closely, and secondly by introducing a third set of beams to produce a triangular grid.

Because of its complexity a triangular grid is not usually an economic system below about 15 m span and can go up to about 24 m. The difference in efficiency between the rectangular grid and the diagonal and triangular grids is shown by the fact that the depth/span ratio for the former is one to thirty whereas for the latter it is one to forty. Since it is very stiff it is well suited to roof construction where the loads are small, the spans are large and deflection needs to be limited.

The structure can be built with universal steel beams welded together or can be of reinforced concrete cast *in situ*, or of pre-cast concrete units which are pre-stressed by wires passed through them and tensioned once the units are in position.

A rectangular, diagonal or triangular grid can also be formed using lattice girders, which gives a lighter roof, and even possibly greater spans. This method of construction should really be considered as a double flat grid, all the top members of the lattice girders forming one grid and all the bottom members forming the other with the two grids connected by the lattice (see Fig. 15.4*b*). It is obviously more complex than a single grid and therefore only becomes economic for roofs over about 20 m span, but can go up to 90 m span with a span/depth ratio of about one-twentyfifth.

Simplicity of jointing techniques is a crucial part of the design as many joints have to be made and it should be arranged so that as much off-site prefabrication as possible can be done.

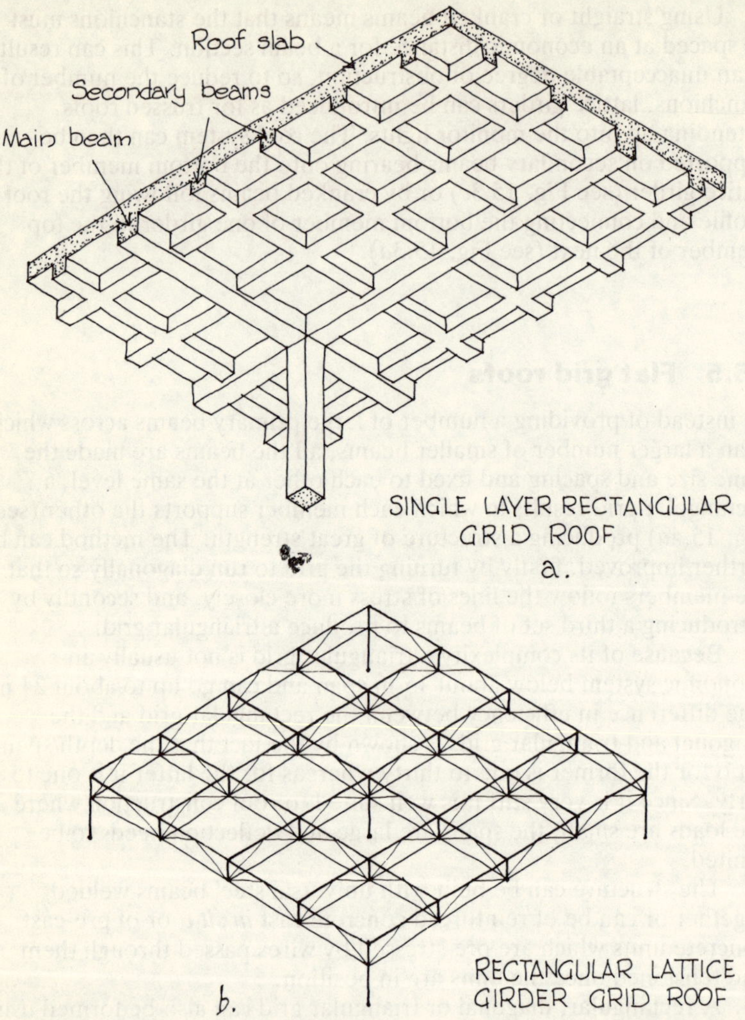

Fig. 15.4 Flat grid roofs

15.6 Space frames

In the last example, the members of the upper and lower grids were
perpendicular to each other. A space frame is also a double layer grid
but the upper members are placed over a line midway between the
lower members so that the lattice bracing lays over at an angle and
connects one bottom member to two top members and vice-versa (see
Fig. 15.5a).

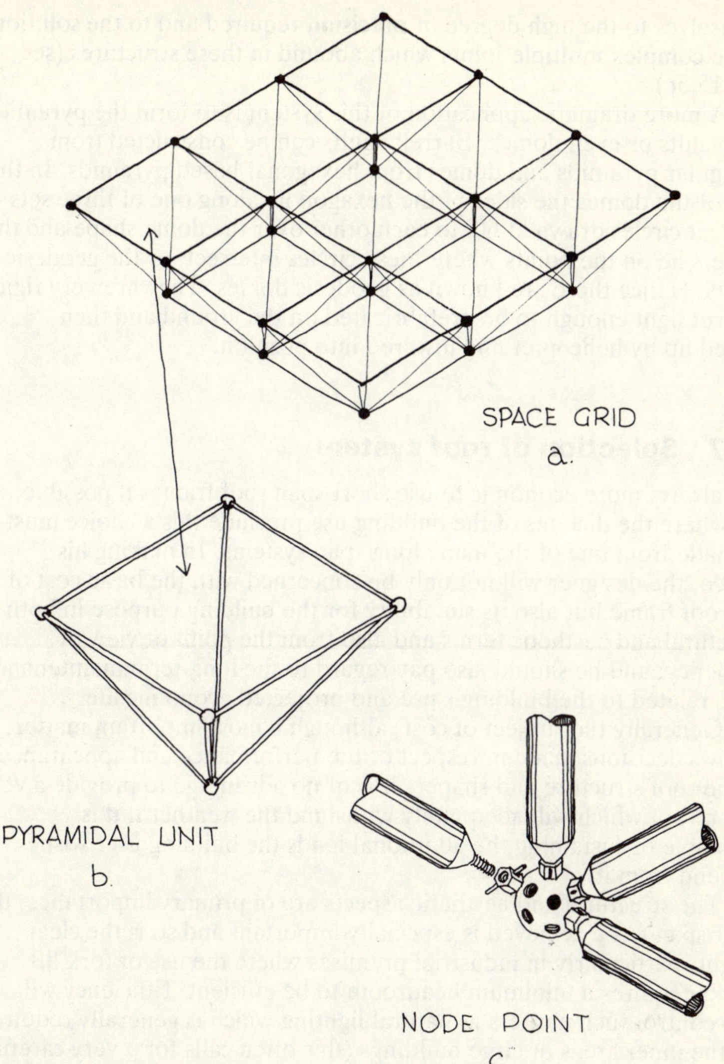

Fig. 15.5 Space frames

Another way of analysing a space frame is to consider it as made up from a series of inverted pyramids with their apices connected (see Fig. 15.5b). The pyramids shown in this illustration have a square base but triangular or hexagonal based pyramids can be used; these produce a more rigid structure.

The material most commonly used is metal, either steel or aluminium alloy in the form of square or circular tubes. These lend themselves to the high degree of precision required and to the solution of the complex multiple joints which abound in these structures (see Fig. 15.5*c*).

A more dramatic application of this system is to form the pyramids into vaults or even domes. Barrell vaults can be constructed from triangular pyramids and domes from hexagonal-based pyramids. In the case of the domes the sides of the hexagon lie along one of three sets of great circles drawn at 60° to each other over the dome shape and the corners lie on the points where these circles intersect, or the geodesic points. Hence these are known as geodesic domes. They are very rigid and yet light enough to be prefabricated on the ground and then picked up by helicopter and lowered into position.

15.7 Selection of roof system

It is always more economic to use short span roof frames if possible, but where the dictates of the building use preclude this a choice must be made from one of the many long span systems. In making his choice, the designer will not only be concerned with the basic cost of the roof frame but also its suitability for the building purpose in both structural and aesthetic terms and also from the point of view of efficiency; and he should also pay regard to the long-term maintenance costs related to the building's use and projected economic life.

Generally the subject of cost, although a most important matter, follows decisions made in respect of the performance and appearance of the roof structure and shape – it is of no advantage to provide a very cheap roof which will adequately withstand the weather if it is incapable of sustaining the additional loads the building user wishes to suspend from it.

The structural and aesthetic aspects are of primary importance; the clear span to be achieved is especially important and so is the clear height, particularly in industrial premises where the use of fork lift trucks requires a minimum headroom to be efficient. Efficiency will also control such matters as natural lighting which is generally required for the inner areas of large buildings (this often calls for a very careful judgement because the glass which lets in the natural light can let out more energy in the form of heat than artificial lights would consume if the natural light was lost by the glass being replaced by a heat insulating structure).

Aesthetics and maintenance are closely related: if appearance is an important consideration then so also must be the maintenance of that appearance to preserve it at its original standard. However, the recurring cost of maintenance against initial cost must be looked at with care for some of these structures: i.e. steel space frame would be cheaper than the same thing in aluminium alloy but would require painting at frequent intervals and with structures of the complexity shown in Fig. 15.5 the cost of painting the multitude of parts can be very high.

The economic life of a building is, in part, related to the ease with which that building can be adapted, altered or extended. These frequently require changes in the roof and hence a structure for buildings where this can be anticipated should lend itself to change. A prestressed concrete geodesic dome, for instance, is as it is and there is no way in which it can be changed, but steel trusses can be extended, cut, welded and generally modified with ease as changing circumstances demand.

Chapter 16

Concrete finishes

16.1 Reasons for surface finishes

When concrete is cast in an ordinary mould, whether *in situ* or
pre-cast, the result is an element of excellent physical properties but
most uninteresting appearance. The colour is dull and often patchy,
the surface is smooth, possibly with fine crazing or pockmarked by
trapped air bubbles, and the whole effect is generally unacceptable.
This arises because although concrete contains natural materials which
possess an interest of their own, the surface of the casting is a cement
paste which completely obscures any natural beauty in the aggregate.

If left, the cement in the cement paste on the surface will weather
away fairly quickly (cement itself is a relatively weak material) to
expose the sand particles and further weathering will remove the sand
to reveal the coarse aggregate. In time, then, the unrelieved dull grey
face of ordinary concrete will improve but this is a slow process of
great variability depending on many factors such as degree of
exposure, type of atmosphere, micro-climate, etc. and is unpredictable
in its outcome. To achieve control over the building's appearance (at
least when newly built) a wide variety of finishing techniques can be
applied either to obtain a specific architectural effect or simply to
obscure any surface crazing which may occur.

16.2 Types of finish

The two objectionable features of cast concrete, mentioned above, are its dull colour and its lack of surface interest. The first can be improved by a deliberate selection of the coarse and fine aggregate on the basis of their natural colour or by the introduction of a colouring medium.

This may be all that is required but if an improvement in the surface interest is desired instead or as well as a change in the colour, many methods can be adopted. The first of these is to change the thing that produces the uninteresting flat surface – the face of the mould against which the concrete is cast – and substitute a surface which forms a texture on the concrete.

A variation of this is to work the surface of the concrete before it has set in the mould to produce a suitable texture. Another variation is to produce a texture so large that it becomes a profiled surface with very powerful architectural effect.

A different approach is to hasten the weathering process whereby the beauty of the natural colours of the aggregate is exposed, by removing the cement paste either by chemical means or mechanically or by a combination of the two. The removal can be carried out shortly after the cement has set and before it has gone hard or after it has gone hard, but preferably not later than three days after casting.

A final alternative is just to accept that concrete is not an attractive material and apply a finishing material which produces the desired surface. This finish may be applied during casting by lining the mould with the finish – sometimes referred to as the transferred aggregate method whereby a selected aggregate is stuck to the mould and then transferred to the concrete by being bonded into the mix – or by covering the surface after the unit has been cast and built in. This solution to the problem, however, does not really qualify as a concrete surface treatment and is therefore outside the scope of this chapter.

16.3 Colour

The colour of a concrete surface as moulded is the product of the combination of the cement colour and the sand colour. If the grey of ordinary Portland cement is combined with the rather pale yellow of many of the common sands available the result is the familiar colour of concrete. This can be considerably lightened by the use of white cement and still further lightened to what is virtually white by the use of silver sand with the white cement.

Other colours can be achieved by the introduction of specially prepared pigments. These should comply with BS 1014 and must be added and thoroughly mixed under strict control, preferably in a mill in a large enough batch for the work to avoid variations in the result, which can easily arise either through an uneven distribution of the pigment throughout the mix or because of slight but significant variations in the proportions of different mixings. Dry ready mixed mortars and concrete which are accurately measured and mixed are available to overcome the problem of inconsistency. If pigments are used the cement content of the mix should be increased by approximately 10 per cent to make up for the loss of strength caused by these additives.

If the cement paste is removed by brushing or prevented from forming then the colour of the coarse aggregate and the sand become much more important. This is dealt with more fully in section 16.7.

16.4 Formed textures

It is extremely difficult to obtain an acceptable surface on a concrete element if it is cast against a smooth flat mould face because of colour variations, fine hair cracks due to shrinkage, pin-holes formed by trapped air and inconsistency in the ease with which the mould is released leading to some areas being rough, etc. To camouflage these defects, the mould faces are lined with a variety of materials which, when removed, impart a texture to the face.

The material used for this purpose must be easily formed into the shape required, waterproof and resistant to attack by cement. Many materials satisfy these requirements, in particular rubber, PVC, thermoplastics and glass-reinforced plastics. All of these require support by a mould but another, quite common, technique does not: using the timber of the mould or formwork itself to impart a texture. This can be seen in many buildings, particularly where the concrete has been cast *in situ*, and the moulds have been made up with sawn boards, usually of narrow width, which leave behind the pattern of the grain.

16.5 Worked textures

A concrete panel can be cast either 'face-down', in which case the finish imprinted by the bottom of the mould is the intended exposed face, or it can be cast 'face-up' when the upper exposed face of the concrete is to be the one shown. This latter method then presents an opportunity to create a texture on this face by hand before the concrete has set.

The simplest version of this is tamping the wet concrete with a board across the mould to produce a series of ridges. Differences in the way this tamping is done will produce different textures and each panel will vary slightly from its neighbour. Further working can be applied using wood or steel floats or an edging tool to produce, say, a smooth margin and rounded arris to a textured panel.

This technique is limited only by the creativity of the designer and the skill and speed of the craftsman in carrying out the design before the mix has set.

16.6 Profiled surfaces

The texturing of panels or *in situ* units can be taken further to achieve
a deliberate profile which produces strong lines across the face, exerting
a considerable influence on the architectural design.

Precast units would be cast 'face-down' into a mould with an
appropriately profiled bottom frequently of metal and generally left
'as-cast' although it is possible to remove the cement paste to reveal
the aggregate if so desired by the designer.

A variation on this is to cast the panel with a ribbed profile and
then, after the panel has been removed from the mould, the top edge
of the rib is knocked off by hand with a hammer or by a percussion
tool with a spade bit to expose the aggregate along the lines of the ribs
but leaving the as-cast surface in the troughs. This results in an
interesting striated finish.

A similar finish can be achieved with *in situ* work where lengths of
rope are lightly attached to the inside face of the formwork before
casting the concrete. When the formwork is removed, the rope remains
embedded in the concrete. After a carefully determined number of
hours the ropes are pulled away, leaving a series of ribs with exposed
aggregate edges and troughs exhibiting the pattern of the rope.

In both these ribbed profiles the rib width has to be restricted to
about 25 mm to ensure that it breaks off easily.

16.7 Exposed aggregate

The easiest, cheapest, often most decorative and therefore most
common means of treating a concrete surface is to expose the coarse
aggregate. The aggregate in this case may be the material of which the
whole mass of concrete is composed or else a carefully selected layer of
aggregate at the surface.

The means by which the aggregate is exposed, the extent to which
it is done and the size, shape and colour of the stone used will mainly
determine the resultant effect. The colour may also be determined to a
varying extent depending on the coarseness of the finish by the choice
of ordinary, white or pigmented cement.

A wide choice of gravels and crushed rock aggregates present the
designer with a range of colours from white to nearly black. The
easiest to achieve is a buff colour since this is derived from the gravels
which occur in many parts of the country. Limestones, calcites and
calcined flints give a white or pink aggregate, the purest white being
obtained from burnt blue silica flints. Pinks and reds can also be
obtained by using granite from Leicestershire, Scotland or Cornwall or
limestone from Devon. Greys and blacks are produced with dolerites,
basalts and whinstone rocks although the Shropshire dolerite produces

a green colour, as also does an aggregate of Westmorland slate or shingle. The use of imported stone, especially crushed marble, further extends this range.

The most common method employed to expose the aggregate is brushing and washing. The freshly cast concrete face is carefully brushed with a wire brush, starting at the bottom of vertical surfaces and the edges of horizontal surfaces, so as to loosen the cement/sand matrix to the desired level and then flushing away the dust and debris with a fine water spray. It is important to maintain consistency in the amount of matrix removed since the greater this is the greater is the amount of stone exposed which will determine the appearance of the concrete surface. Retarders may be painted on the face of the mould to delay the set at the surface and assist this.

Brushing and washing should be carried out within two to six hours of casting the concrete but other methods can be used on mature concrete, among these being abrasive blasting of the surface to erode the matrix. This is performed either with a jet of grit in compressed air, or sand in a jet of water, or lead shot which is ejected through a special head which collects and recirculates the shot. The process is attractive in that it can be carried out at any time, although within three days of casting is preferable; it also creates opportunities for added refinement of decoration by varying the depth to which the erosion is taken in controlled areas or even leaving some areas untouched by the aid of a mask fixed to the concrete face.

Another process carried out on mature concrete is to grind the surface down to expose the aggregate and then follow with successively finer grinding discs until a polish is achieved. This can be used on both flat and profiled surfaces and when combined with carefully selected aggregates and cement can produce a very high quality finish.

Aggregates carefully selected for their appearance may not be the same as the aggregate that would be chosen for the mass of the concrete either because of inherent strength or, more generally, because of cost. In this case, the aggregate for the face would be mixed and laid in the bottom of the mould followed by the backing concrete, or a bed of sand would be laid in the mould, into which the dry aggregate is bedded, again followed by the backing concrete. After removing the concrete panel the sand is brushed away to reveal the aggregate embedded in the concrete. This latter process is also used where the size of the aggregate exceeds 60 mm and is hand-placed one stone at a time in the sand.

If the concrete is cast face up, the selected aggregate can be rolled or trowelled into the face before the concrete sets or into a specially mixed and tinted mortar on the concrete face. This technique produces an evenly distributed and dense layer of aggregate.

Tooled finishes can also be worked on concrete by bush hammering or point tooling. Almost any aggregate can be treated this way but it is most successful on stones which cut rather than shatter, such as limestones. The process is to work over the surface with a powered percussion tool with either a hammer/or a pointed bit so as to fracture the surface to a medium roughness. It is necessary to have a substantial robust concrete unit to work on and the working cannot be taken nearer than 20 mm from any square arrises without danger of damage occurring due to spalling.

Part IV

Internal construction

Chapter 17

Suspended floors

17.1 Pre-formed floors generally

Many pre-formed floors have one outstanding advantage over floors formed in position and that is that they are available for use as soon as they are laid. An exception to this is the concrete floor system embodying a structural topping. Another advantage is that they do not require any temporary support, thus leaving the floor below free from obstructions.

The choice of material and method is dictated by span, loading, fire resistance and sound insulation properties. Short span lightly loaded floors are most economically formed in timber either in prefabricated panels or cut and assembled *in situ*. Timber floors, however, do not offer very much fire resistance or sound insulation. If the loading exceeds 4.00 kN/m^2, or the span exceeds 5 m, or if high fire resistance or sound insulation is required, timber is not an economic answer; the material usually chosen is concrete. This, in all the methods of construction used, will span from 3.0 m to 6.0 m with maximum economy, and up to 12 m if prestressing techniques are employed.

Steel floors, of the open mesh type described in this chapter, have a particular range of application and are limited in span to about 3.0 m but in combination with concrete can be used economically for spans up to 8.0 m.

17.2   Pre-cast concrete floors

There are two ways of solving the problem of pre-casting a concrete
floor. The overriding difficulties are weight and size: a complete floor
panel is inconveniently large to maneouvre and excessively heavy to
handle in one piece. Pre-cast systems break this panel down into
smaller component pieces, either by dividing it lengthwise into strips of
slab or beams which are laid across the bearings; the top is then tidied
up with a screed. Or else the division is made horizontally in the depth
of the slab, the lower part being pre-cast and the upper part formed on
site as a structural screed. The structural screed not only provides the
finished floor surface but, by being tied to the bottom half, forms the
compression zone of the slab structure.

17.3   Pre-cast concrete floor panels

The maximum size of any building element is controlled by the on-site
provision for handling but, generally, the larger the element the less it
costs to fix. The economic balance between size and handling costs
always must be considered with pre-cast concrete. Where a crane is
already required on site it makes sense to use it to its maximum,
therefore the elements of the building can be much larger units. This is
the basis for selecting pre-cast floor panels.

The size of floor panels is mostly between 1.2 and 2.0 m wide but
larger units are possible where it is feasible to use them. They can be
of either the type which is the whole slab slit lengthwise into panels or
the type which is the tensile area of the floor only, requiring a
structural screed to finish off the construction.

Tee-beam, ribbed or finned panels (Fig. 17.1*a*) are complete floor
units requiring only a finishing screed. By concentrating the tensile
reinforcement into ribs the effective depth of the slab is retained but
much weight is saved thus making even larger units possible. In many
buildings it is necessary to fix a ceiling to the bottom edges of the ribs
to give a smooth soffite and by doing so service ducts are formed along
the length of the slab but, unfortunately, they are not linked by a cross
duct, unless special provision is made across their ends, or the ceiling is
suspended below the line of the ribs.

Another wide slab or panel which only requires a finishing screed
is a multi-core slab which has continuous voids formed within the panel
to reduce the weight. This has the advantage that the soffite is closed,
making a ceiling finish less expensive but, at the same time, eliminating
the ducts formed by the ribbed panels (see Fig. 17.1*b*).

The type of panel which forms the tensile zone and requires an *in
situ* structural topping is usually a prestressed concrete plank 50 mm

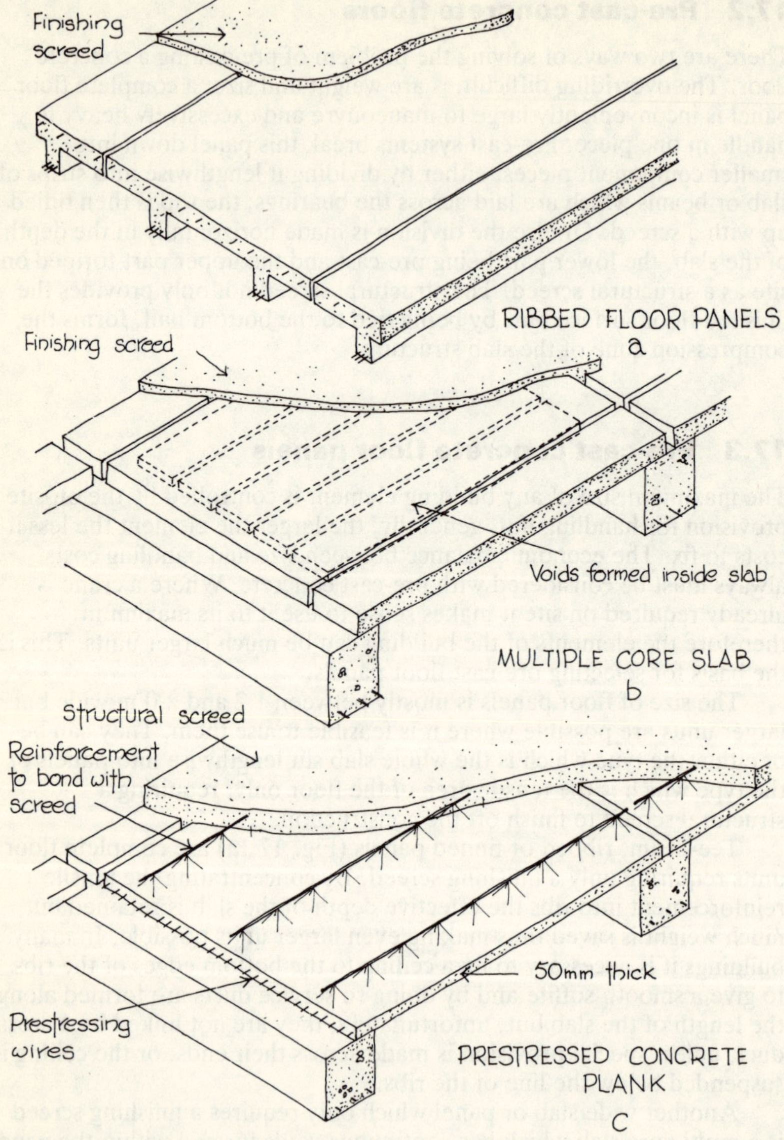

Fig. 17.1 Concrete floor panels

thick with projecting stirrups and bars provided to bond with the
screed. Typical details are shown in Fig. 17.1c. These panels, being
very thin, are also relatively light and can be produced up to 2.4 m
wide.

17.4 Pre-cast beam floors

Where very heavy lifting gear capable of handling the large floor panels described in section 17.3 is not available a smaller unit must be used. One of the solutions is to reduce the width of the floor elements until they are of the size of beams, fix them side by side and finish with a screed. To effect further savings in weight and materials, these beams are formed with an 'I' or channel section or are made hollow. All three versions are shown in Fig. 17.2. The 'I' section beams are suitable for

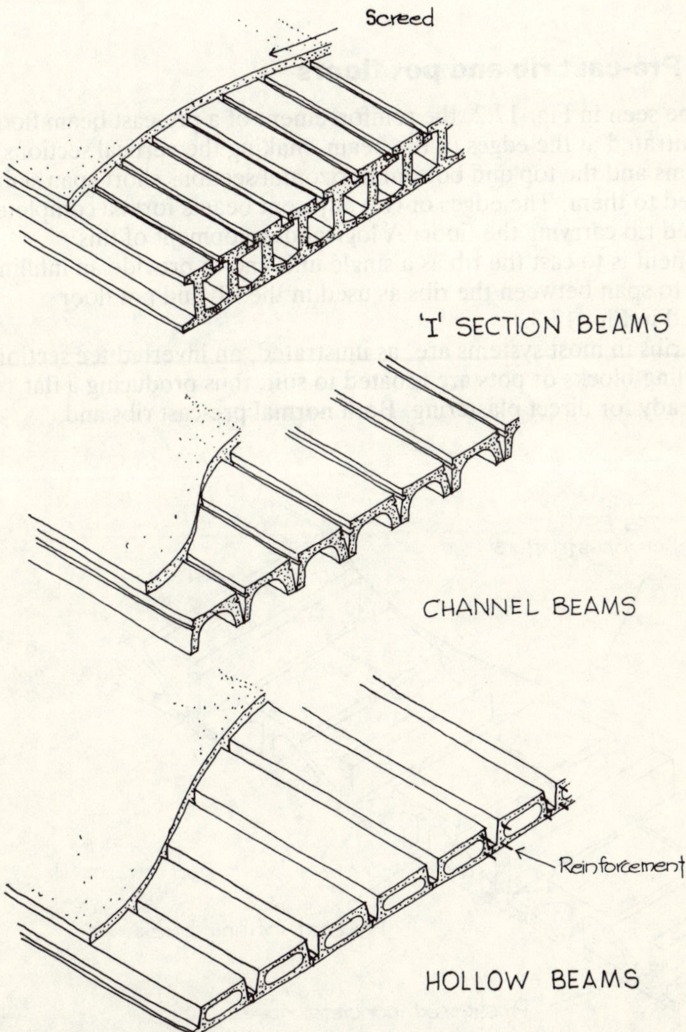

Fig. 17.2 Pre-cast concrete beam floors

prestressing which, despite the higher cost of the better quality concrete required and the extra production expenses, can prove to be a more economic answer in certain situations.

The channel beam floor, like the ribbed panel, requires a ceiling system to provide a flat soffite, but produces ducts for services running the length of the span only.

The hollow beam floor produces a continuous soffite which, if the beams are accurately made and carefully installed, only requires a plaster skim as a finish.

17.5 Pre-cast rib and pot floors

As can be seen in Fig. 17.2, the reinforcement of a pre-cast beam floor is concentrated at the edges of the beam, making the vertical sections thin beams and the top and bottom horizontal sections short span slabs connected to them. The edges of two adjacent beams form a complete reinforced rib carrying the floor. A logical development of this arrangement is to cast the rib as a single unit and to provide an infilling element to span between the ribs as used in the rib and pot floor illustrated in Fig. 17.3.

The ribs in most systems are, as illustrated, an inverted tee section. The infilling blocks or pots are rebated to suit, thus producing a flat soffite ready for direct plastering. Both normal pre-cast ribs and

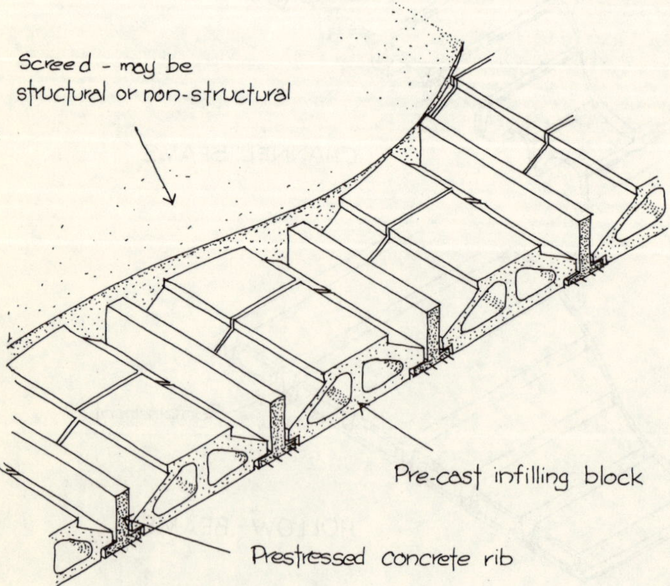

Fig. 17.3 Typical rib and pot floors

prestressed ribs are used, and the infilling units can be hollow pre-cast concrete blocks, hollow clay tiles, pots, lightweight concrete or woodwool slabs.

The system produces a very light form of pre-cast floor which is very easy to install and, especially when prestressed, a very economical system because the concrete – high grade in the case of prestressed – is confined to the small ribs, the majority of the floor structure being the light and cheap infilling units.

17.6 Steel floors

As a material on its own, steel does not lend itself to the formation of complete floors in buildings, although the various decks of a modern ship are all satisfactorily made of steel plates probably because a modern shipwright's firm is better equipped to work and handle steel plate than is a builder.

It is sometimes used in combination with concrete to form a composite floor (see section 17.9) but purely on their own steel floors are confined to access walkways, platforms, storage galleries, the landings to fire escape stairs and similar solely functional purposes where appearance and noise control are not important and the floor is not required to offer a fire resistant membrane.

17.7 Open steel floors

Steel sheet is not only very strong but also very heavy in the sizes required for a floor. To reduce this weight these floors are usually constructed as an open arrangement of bars or strips, or made up from expanded metal fixed to suitable steel angle bearers. Figure 17.4 shows two types of open steel floor.

These are generally to be found in industrial premises and have the advantage that although they are strong enough to walk on in safety, they permit the passage of light and air to the areas below the floor.

17.8 Steel panel floors

One way in which steel is used for a floor is to make it up into small panels which are supported in steel stringers resting on steel pedestals, or short legs, fixed to a solid sub-floor. This system, as shown in Fig. 17.4, finds a particular application in offices and computer rooms, where complicated sub-floor wiring is required. The steel panels in offices are in the form of trays filled with chip board to which a variety of floorings can be fixed. The panels simply rest on the stringers and offer a quick and easy means of access to the services below.

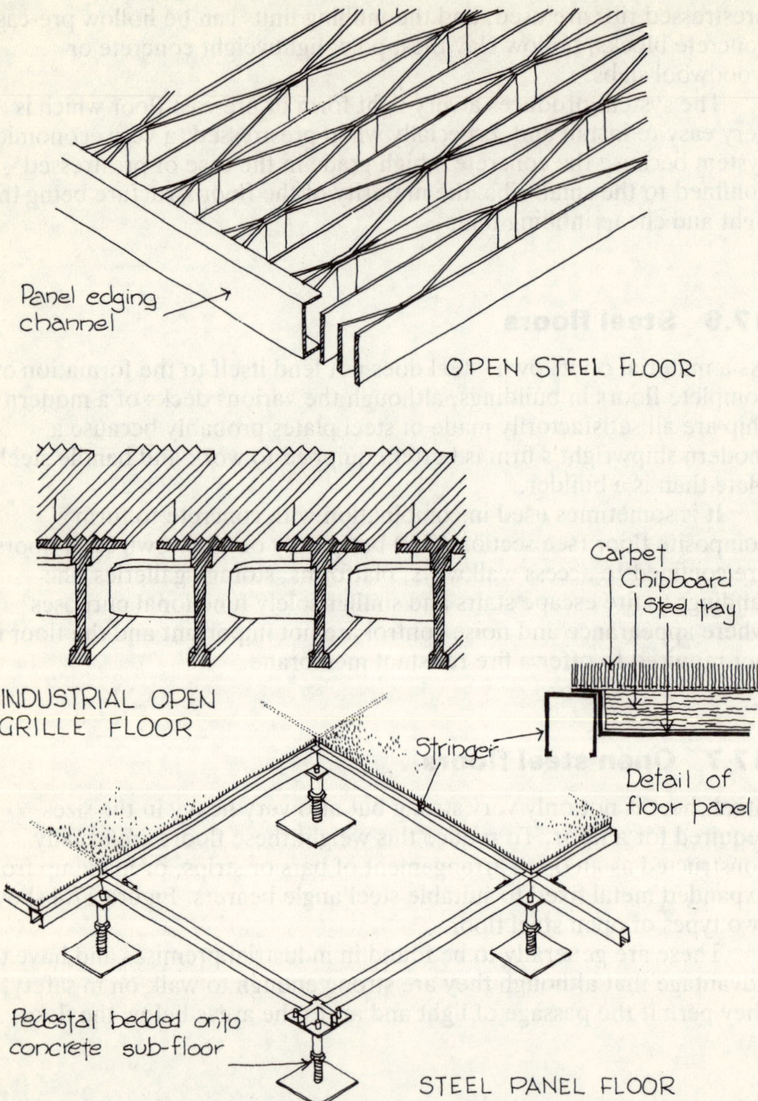

Fig. 17.4 Steel floors

17.9 Composite floors

By combining the most suitable materials and using both on- and
off-site working to simplify installation, the best economic solution to
the design of a floor can often be achieved. The combinations used in
the main are either steel and concrete, or steel and timber.

17.9.1 Steel and concrete floors

An old form of composite floor is the filler joist floor where small steel joists at closely spaced centres combine with an *in situ* concrete floor cast round them, frequently with an arched soffite as shown in Fig. 17.5. To reduce dead weight, the builders in the past used coke breeze as the aggregate for the concrete. Unfortunately coke breeze contains a certain amount of sulphur and sulphur salts which combine with moisture to form acids which in turn attack the steel of the filler joists, leading to their gradual deterioration.

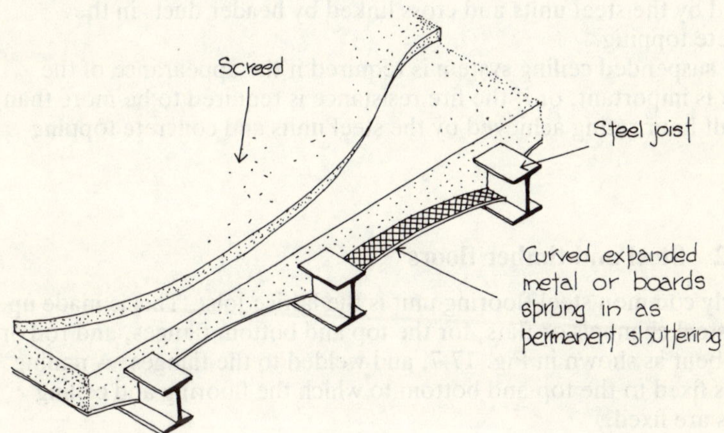

Fig. 17.5   Filler joist floors

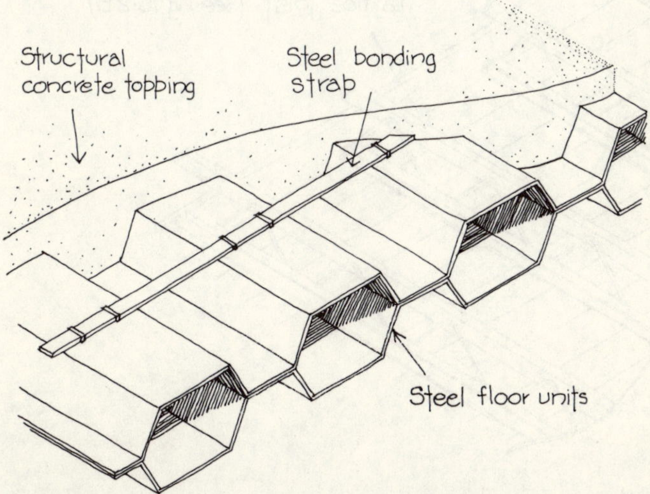

Fig. 17.6   Cellular steel floors

A newer form of steel and concrete floor is composed of a cellular steel structure with a structural concrete topping as shown in Fig. 17.6. The steel cell units are made in a range of depths to suit the span required and are linked by a steel strap at intervals, which also serves to bond the concrete to the steel so that the two act together.

The system is light in comparison to pre-cast concrete beams but possesses the same advantage of providing a useable working surface as soon as the units are fixed in position – even before the concrete is placed – and does not require any supports from the floor below.

Service cables can be threaded longitudinally through the ducts formed by the steel units and cross linked by header ducts in the concrete topping.

A suspended ceiling system is required if the appearance of the soffite is important, or if the fire resistance is required to be more than the half hour rating achieved by the steel units and concrete topping alone.

17.9.2 Steel and timber floors

A fairly common steel flooring unit is the lattice joist. This is made up from steel channels or flats, for the top and bottom flanges, and rod or tube, bent as shown in Fig. 17.7, and welded to the flanges. A nailing strip is fixed to the top and bottom to which the flooring and ceiling panels are fixed.

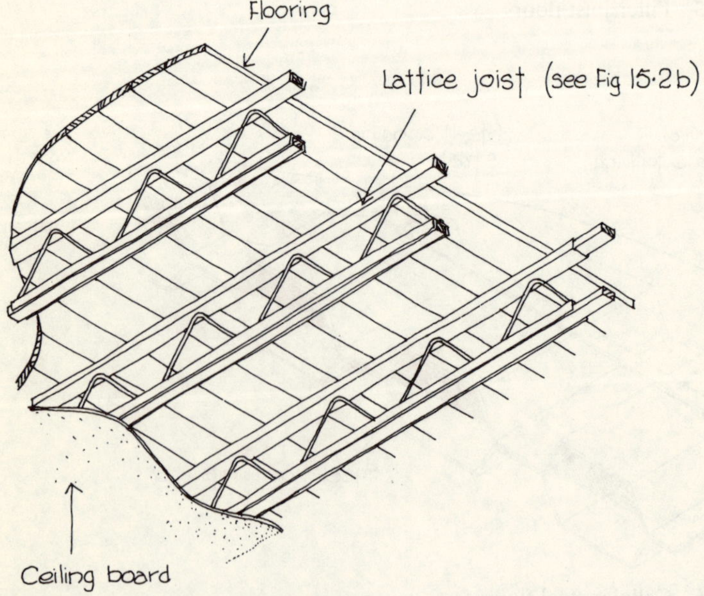

Fig. 17.7 Steel and timber floors

These lattice joists are much stronger than solid timber joists, and produce a thinner floor, but they also allow great freedom in the running of services in the floor thickness. They can be spaced at the normal 400 mm centres to support 19 mm thick flooring where domestic loads are involved; but greater economy is usually achieved by spacing the joists at 600 mm centres and using 25 mm flooring with a thicker ceiling.

The flooring can be traditional softwood tongued and grooved boarding; but where a floor finish is to be laid over it, the preference is for pre-formed, tongued and grooved flooring grade chipboard panels, 1.2 m square. These are screwed down and produce a much better surface to which to fix flooring.

The soffite can be finished either with plasterboard fixed in the normal way to the bottom nailing strip or else with pre-formed ceiling panels fixed in the same way as plasterboard or supported by a suspended ceiling grid.

This system is more appropriate for domestic work than any other, particularly where a large development is planned when bulk ordering brings the cost down to a competitive level.

Chapter 18

Prefabricated stairs

18.1 Advantages

The advantages of installing prefabricated concrete or steel stairs over those formed *in situ* follow the same pattern as any other prefabricated units; that is, speed of erection, immediate availability of use, financial economies through industrialised off-site production, and greater accuracy and consistency. Equally the same constraints apply such as the need to standardise the elements to take maximum advantage of the factory production process; the careful control of dimensions so as to ensure that the on-site work remains within the specified tolerances; the necessity to consider the design of the elements with respect to transport to the site; handling on site and locating in position.

Probably the greatest asset possessed by prefabricated stairs is availability for use as soon as they are fixed. This is equally true whether the units are as small as single steps where the workman can stand on the lower ones to fix the upper ones, or whether they are made up in complete flights, in which case their installation replaces the ladder which probably preceded the finished stairs in the stairwell, eliminating the loss of access to just an hour or two.

18.2 Concrete stairs

Except for small works and domestic installations, concrete is the most common material used for stairs because of its fire-resisting properties.

Since most stairs provide a means of escape, their resistance to fire is a subject of considerable importance. In addition to their fire-resisting properties, concrete stairs are strong, quiet and capable of being produced in many different designs either as the designer wishes or as dictated by the structural support available. The following is a selection.

18.3 Straight flight concrete stairs

The restrictions imposed by the Building Regulations on the size of each step and a maximum of sixteen steps in one flight make the pre-casting of a complete flight between landings a feasible proposition from the aspect of transportation and handling, provided that, in the latter case, a crane is also required for other purposes at the same time.

In design this is treated as a simple slab which is fixed in an inclined position and finished with a surface formation to produce the required steps. Typical details are in Fig. 18.1, which shows a stair without strings. This is usually possible because of the short spans generally encountered.

The detail of the joints between the slab and the landings must be carefully considered to achieve a simple installation but yet allow for expansion and contraction.

18.4 Cranked concrete slab stairs

In most building designs the stair construction is a totally separate operation to the rest of the concrete work either because of its place in the programme (if may be left until later to allow loads to be hoisted up the stair well or to prevent damage to the stairs) or because it is of a completely different form of construction to the rest (such as a steel-framed structure with a patent floor system and solid reinforced concrete stairs). If it is a separate operation the straight flight stairs is not a suitable method because the need to cast the landings *in situ* cancels any advantages gained by pre-casting the flights. The obvious answer is to pre-cast the landings as well which is usually the case with a cranked slab stair.

Figure 18.2 shows the formation of this type of stair with part of both the top and bottom landings cast in with the flight as a single slab bent into a rather flattened 'Z' shape and spanning from end to end.

A small quantity of *in situ* concrete may be required cast between the two parts of each landing but this is a negligible operation compared to casting the whole landing.

The ends of the pre-cast sections bear onto specially formed beams provided in the enclosing structure. It therefore follows that the

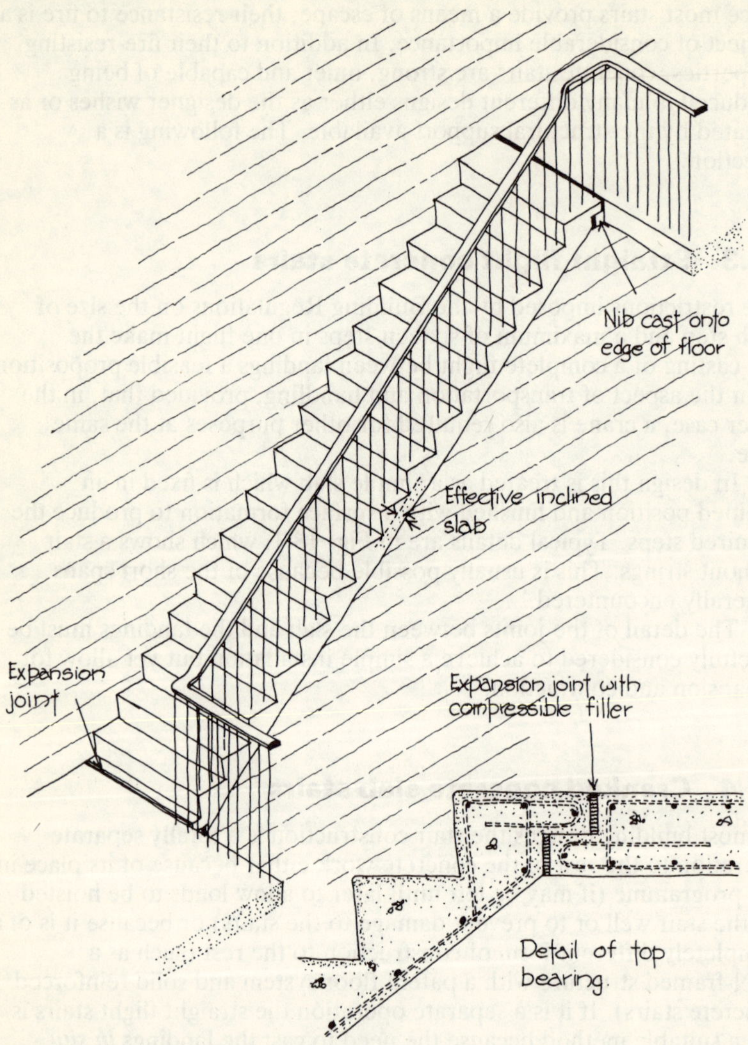

Fig. 18.1 Straight flight pre-cast stair

method is only suitable if that enclosing structure is sufficiently strong – not a light-weight glazed box attached to the outside of the building as can be seen in some designs.

18.5 Concrete spine beam stairs

This is an alternative to a straight flight stairs, lighter both in

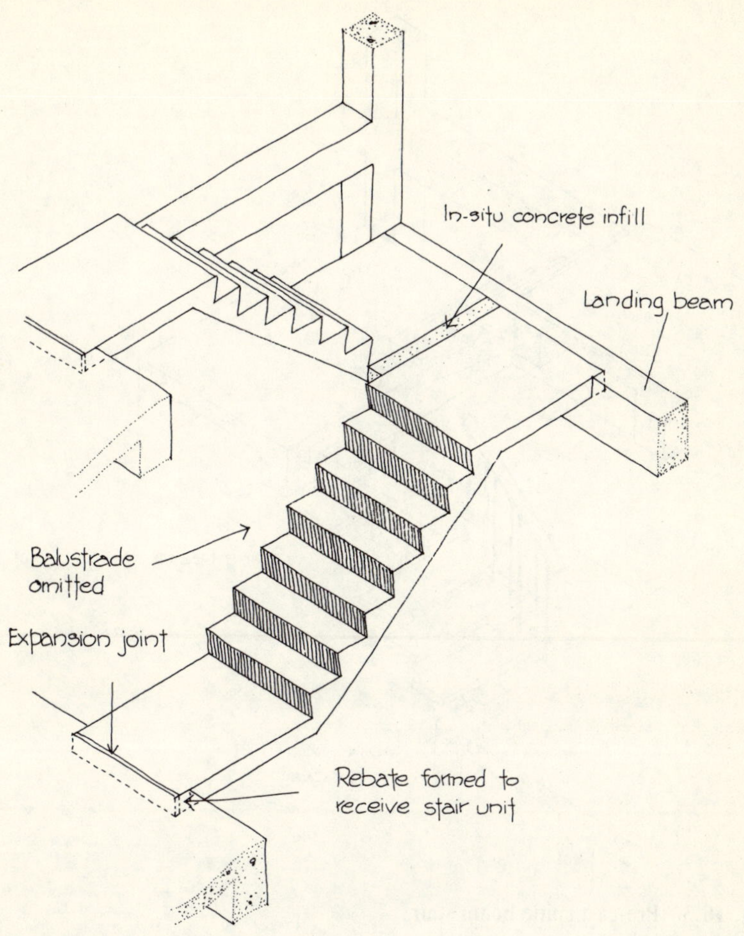

Fig. 18.2   Cranked concrete slab stair

appearance and in fact since it is formed with a single beam placed
centrally under each flight to which are attached cantilevered treads.
Generally risers are omitted either as a design feature or to allow light
to penetrate.

Like the straight flight stairs, the landings are usually cast *in situ*,
with pockets formed to receive the ends of the spine beam, which
means that if it is a dog-leg stair arrangement, support must be
available at mid-height between the main beams for the half landing
(see Fig. 18.3).

This form of stair has two advantages. Firstly, because it is broken
down into smaller pieces each is lighter than the single straight flight
slab and therefore much easier to handle, requiring only simple lifting

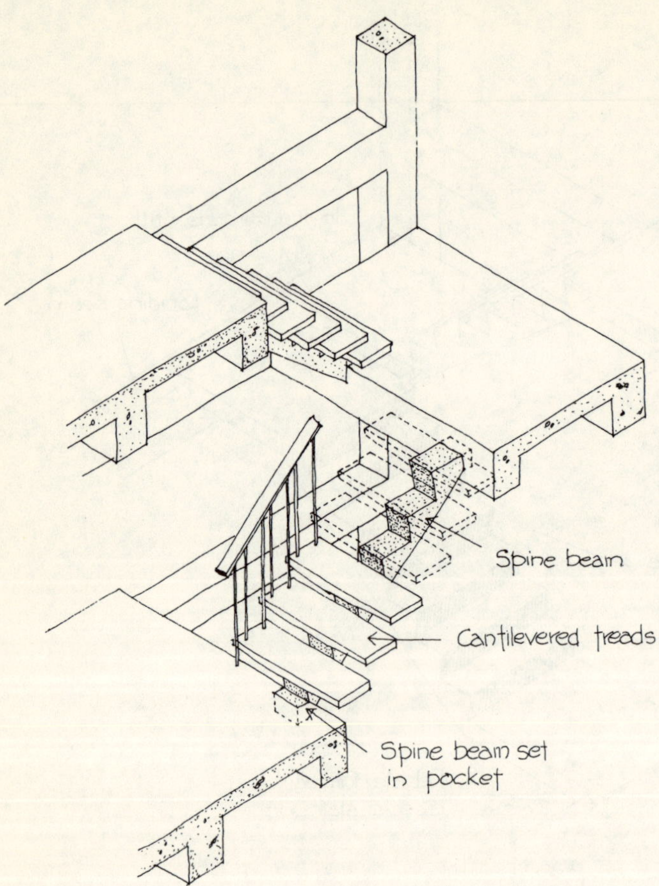

Fig. 18.3 Pre-cast spine beam stair

tackle; and secondly being in separate parts means that other materials can be introduced such as timber cantilevered treads rather than all concrete.

18.6 Concrete stair units

Where the stair well is enclosed in substantial brick walls the stairs can be formed as individual steps, the outer end of each being built into the wall. The step may then either be a simple cantilever or may be formed in the manner of stone steps where the lower edge of the riser is supported by the back edge of the tread of the step below as shown in Fig. 18.4a.

When the necessary walls are available these methods can offer a

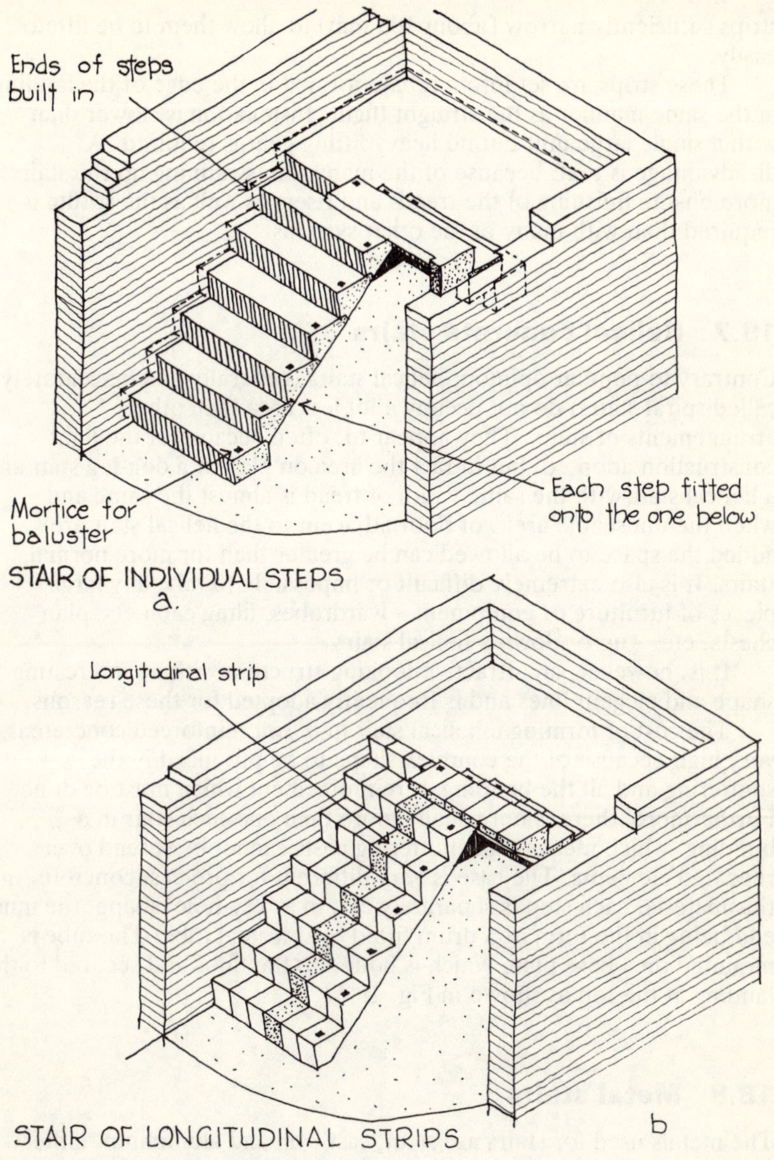

Fig. 18.4   Pre-cast stair units

fairly easily constructed stair as the units are relatively easy to handle and lift but the process is slow and the cantilevered form will require each step to be propped.

Another type of stair unit aimed at reducing the weight is shown in Fig. 18.4*b*. This is simply a straight flight stair cut into longitudinal strips sufficiently narrow (about 100 mm) to allow them to be lifted easily.

These strips are set into a rebate formed in the edge of the landing in the same manner as the straight flight. Installation is slower than with a single large unit but no heavy lifting gear is required. A disadvantage is that, because of the many joints running up the stairs, more on-site finishing of the treads and risers as well as the soffite is required than with many of the other systems.

18.7 Helical concrete stairs

Contrary to popular opinion, helical stairs (generally and inaccurately called spiral stairs) do not occupy a lot less area than other arrangements of stairs. They appear to, often because of the light construction adopted, but in fact the area on plan of a dog-leg stair and a helical stair with the same width of tread is almost the same and when the unuseable areas of floor adjacent to the helical stair are added the space to be allowed can be greater than for more normal stairs. It is also extremely difficult or impossible to take any large pieces of furniture or equipment – wardrobes, filing cabinets, plan chests, etc. – up or down a helical stair.

It is, however, an attractive-looking structure with an interesting shape and elegant lines and is frequently adopted for these reasons.

The cost of forming a helical stair in *in situ* reinforced concrete is very high because of the complex shape to be produced in the shuttering and all the bending of reinforcement which must be done. Furthermore, there is not usually more than one such stair in a building, which means that shuttering costs cannot be spread over repeated elements. The case is very different for pre-cast concrete. In this material each step is separately cast in a 'key-hole' shape, the inner end being in the form of a drum fitted over a steel tube. The tube is mounted on a base plate which is bolted to the floor and secured to the landing at the top as shown in Fig. 18.5.

18.8 Metal stairs

The metals used for stairs are steel, cast iron and aluminium, either singly or in combination – a steel framework with cast iron treads is a common arrangement which takes advantage of the adaptability of steel and the rigidity of cast iron. It is also quite usual to find the hard appearance of the metal softened by timber treads and handrails. Aluminium is more decorative than steel and also retains its original appearance longer but as it is less strong its use tends to be restricted to balustrading.

External fire escape stairs are frequently made of metal because it

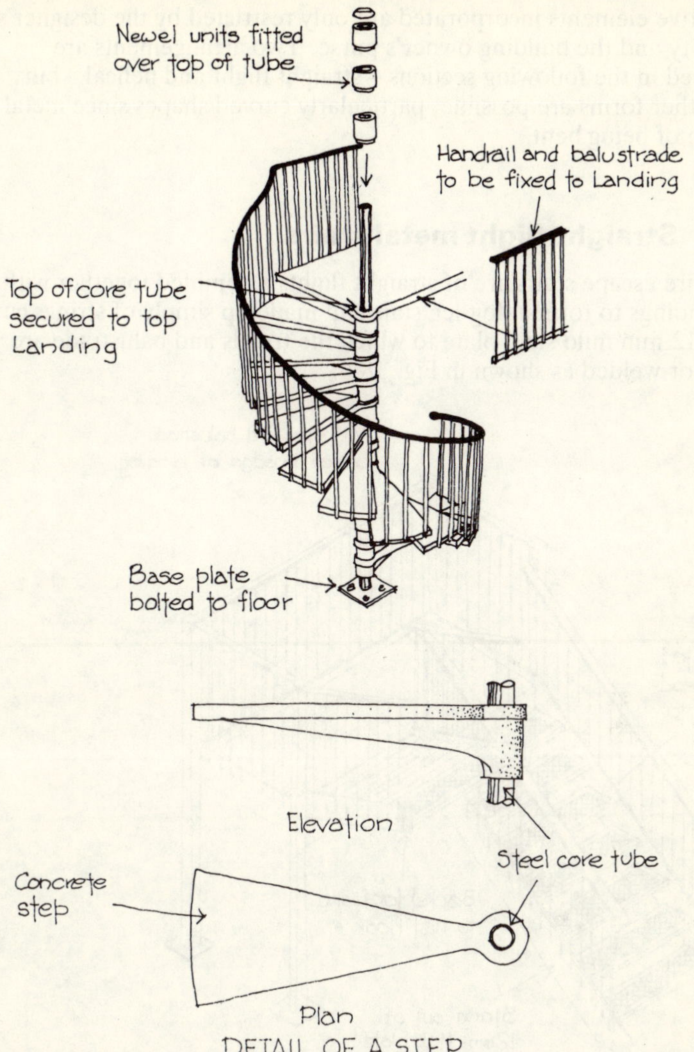

Fig. 18.5 Pre-cast helical stair

is light, easy to erect, relatively cheap and, of course, fire resisting. It is also possible, by perforating the treads, to avoid rain collecting on the stairs leading to the possible hazard of ice formation.

Internally the use of metal for stairs ranges from the basic access, e.g. to upper storage levels in a warehouse, to delicate cast iron helical stairs for use in a house, or to monumental stairs in decorative metal work painted and gilded and capped with exotic hardwoods.

The possible stair arrangements are many and varied and the

decorative elements incorporated are only restricted by the designer's ingenuity and the building owner's purse. Two arrangements are described in the following sections – straight flight and helical – but most other forms are possible, particularly curved shapes since metal is capable of being bent.

18.9   Straight flight metal stairs

Many fire escape stairs are of straight flights assembled together with half landings to form a dog-leg stair and made up simply of strings cut out of 12 mm mild steel plate to which the treads and balustrade are bolted or welded as shown in Fig. 18.6.

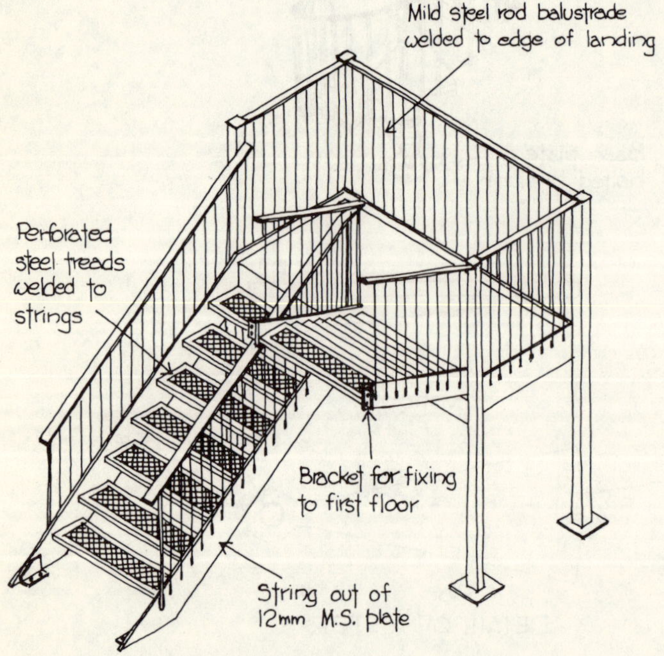

Fig. 18.6   Straight flight steel stair

The more decorative internal stairs may use steel tube or steel box strings to which brackets are welded to support hardwood treads as shown in Fig. 18.7.

In addition to the built-up steel stairs there are several stair systems made up from purpose-formed aluminium alloy sections which clip together. The treads are faced with vinyl or similar finish and a non-slip nosing insert. This system gives a stair which is very easy to

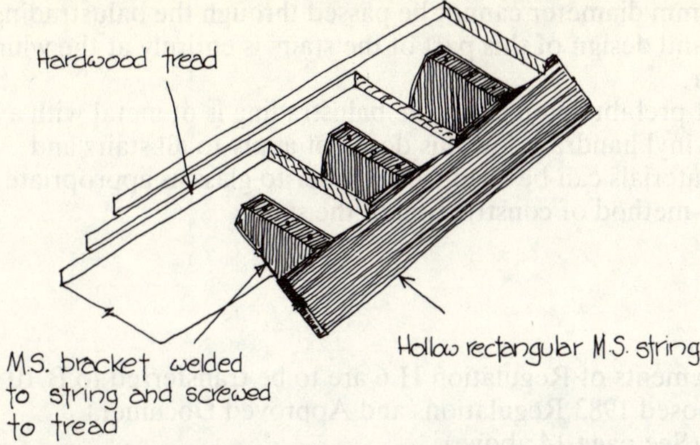

Fig. 18.7   M.S. box string and hardwood tread stair

handle and install in awkward positions where the use is not heavy and fire resistance is not important but where the decorative quality does matter.

18.10   Metal helical stairs

Because it is possible to bend, metal, particularly steel, is a convenient material from which to construct a helical stair. The same forms of construction can be used as illustrated in Figs 18.6 and 18.7 except that the strings are rolled to the required inside and outside curves and the treads are cut on the taper.

In addition to these methods cast iron or pressed steel key-hole shaped steps can be used, fitted over a central core tube or steel rod in the manner of the concrete stair illustrated in Fig. 18.5. The outer ends of the treads are connected by threaded pin on the bottom end of each baluster.

A metal helical stair can be used for fire escape purposes provided that the diameter is at least 1.5 m, the total rise not more than 9.0 m and the number of people who would use it does not exceed 50. Their use however is more common for an internal simple access stair which is not required to be a fire escape route. They are usually quite small in diameter with decorative treads and balustrading.

18.11   Balustrading

All stairs must be guarded by a balustrading and provided with a handrail as laid down in Part H of the Building Regulations.[1] Provided that the handrail is 840 mm above the string and, if appropriate, a sphere of 100 mm diameter cannot be passed through the balustrading, the materials and design of this part of the stairs is entirely at the whim of the designer.

With most prefabricated stairs the balustrading is of metal with a hardwood or vinyl handrail, but this does not apply to all stairs and many other materials can be used from timber to glass as appropriate to the use and method of construction of the stairs.

Note

1. The requirements of Regulation H.6 are to be transferred to B.10 of the proposed 1983 Regulations and Approved Document AD.A.08. (See page 14 above).

Chapter 19

Industrial doors and shutters

19.1 Door types

The design problems associated with industrial doors stem from the fact that they are large and heavy, and resolve into two considerations; how to move whatever is provided to close the opening and where it is to be parked when it has been moved.

The size and weight will preclude the use of normal hinges because of the forces at the hinges and the difficulties of moving and holding the door, especially in a wind. One solution to this is to arrange for the door to slide sideways, either on rollers at the bottom or hanging from track at the top. The problem with this method is that there must be as much wall beside the opening as there is aperture, which is not always the case, and that that wall area must be kept free of any fittings or equipment either fixed on it or even stood against it. This can be restrictive in the use of the industrial premises.

The amount of wall required for parking the doors can be reduced if they are divided into narrow widths, hinged together and supported on pivots in such a way that the door folds into a concertina fold when opened. The folded door can be parked either within the opening, in which case it reduces the useable width of the opening, or beside the opening, when it will occupy a small amount of wall but will also project at right angles to the wall, presenting what may be an unacceptable obstruction.

Numerous variations of the simple sliding or folding door exist, all endeavouring to solve the problem of where to park the doors

184

conveniently. An alternative solution is to remove the structure closing the opening in a vertical direction and thereby leave the floor and adjacent walls free. This vertical operation for industrial purposes resolves into shutter doors or slideover doors (see Fig. 19.1).

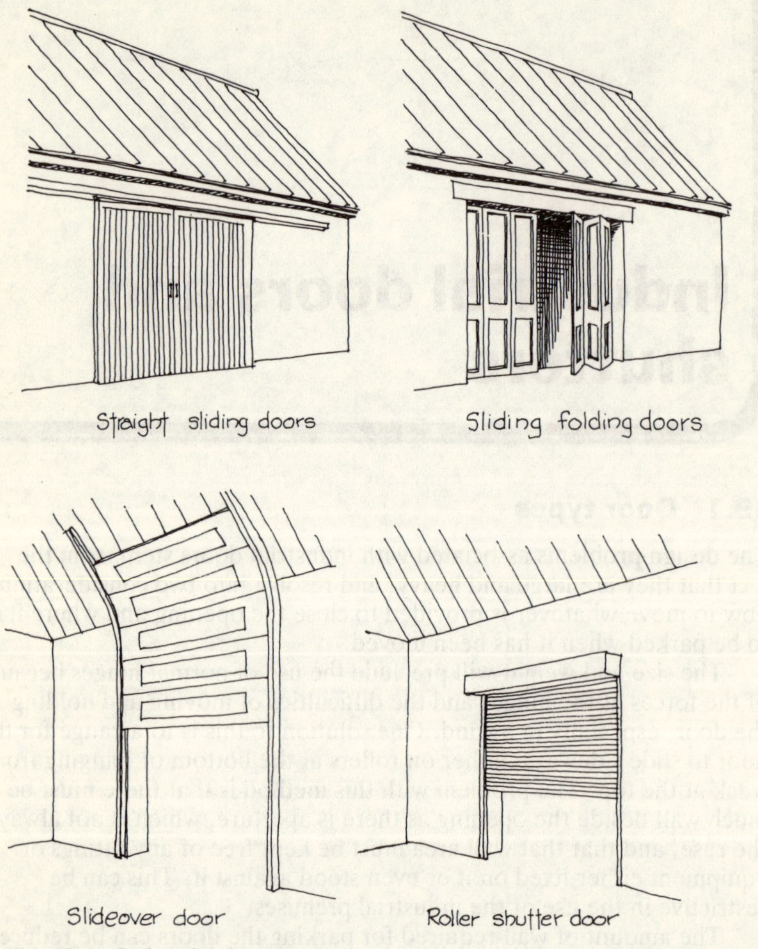

Fig. 19.1 Industrial door types

19.2 Sliding doors

Horizontal sliding doors, as already mentioned, can either be hung from a top track or run on rollers fitted to the bottom edge. With either type, standard door gear is available capable of carrying a door of up to 550 kg and 13 m^2 in area. However, with the top track suspension, the building structure must be capable of supporting the load on the track fixings. Special gear is available to accommodate larger and heavier doors.

Figure 19.2 shows typical sections for top-hung straight sliding doors, which can be fitted to the inside or the outside face of the wall. The outside hung door does not interfere with the use of the building unless it covers up a window or door when open, but the track needs t be protected from the weather by a hood.

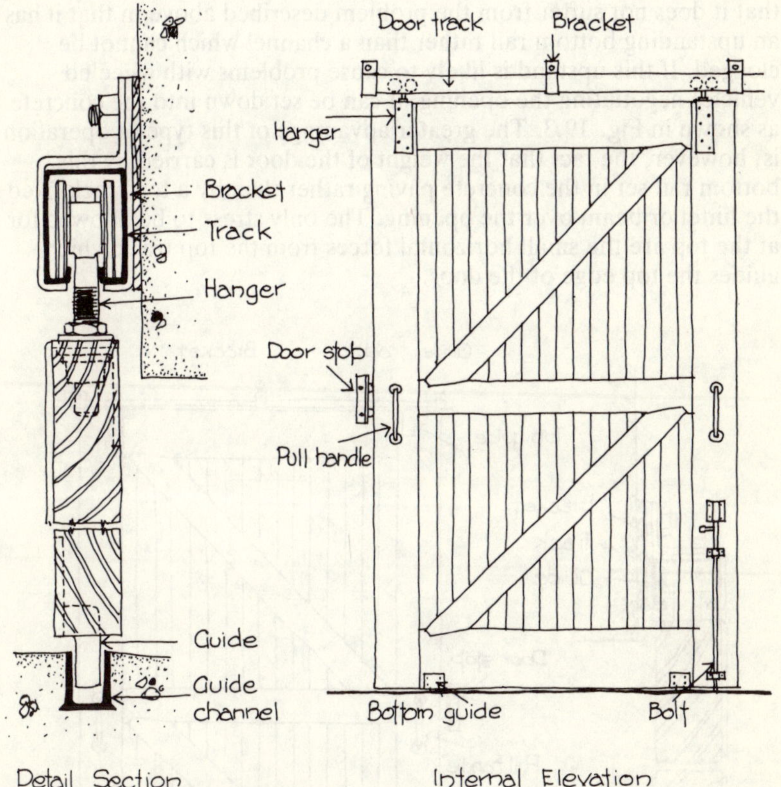

Fig. 19.2 Top hung straight sliding door

The top track consists of channel section galvanised steel up to 4 mm thick fixed to the wall with brackets spaced between 600 and 900 mm apart. Two hangers per door are run in the track, each hanger comprises a four wheeled trolley, a suspension arm and a means of fixing the arm to the door which affords a vertical and lateral adjustment. The type of fixing is determined by whether the door is of timber of steel.

At the bottom the door is retained in position by a pair of bottom guides running in a steel or aluminium channel set in the concrete paving. It is this channel which can be one of the principal causes of trouble. When set flush with the ground or floor, it is constantly being filled with rubbish, dirt and grit which, if not regularly removed, can jam the bottom guides. The bottom guide is usually designed in a plough shape to remove most of the material in the channel as the door is operated but this only deposits it adjacent to the channel from where it is almost immediately pushed or washed back in again.

One advantage of the bottom roller door over the top hung door is that it does not suffer from the problem described above in that it has an upstanding bottom rail rather than a channel which cannot be clogged. If this upstand is likely to cause problems with wheeled vehicles negotiating the opening, it can be set down into the concrete as shown in Fig. 19.3. The greater advantage of this type of operation is, however, the fact that the weight of the door is carried by this bottom rail set in the concrete paving rather than by a top track fixed to the lintel or beam over the opening. The only stress to be allowed for at the top are the small horizontal forces from the top track which guides the top edge of the door.

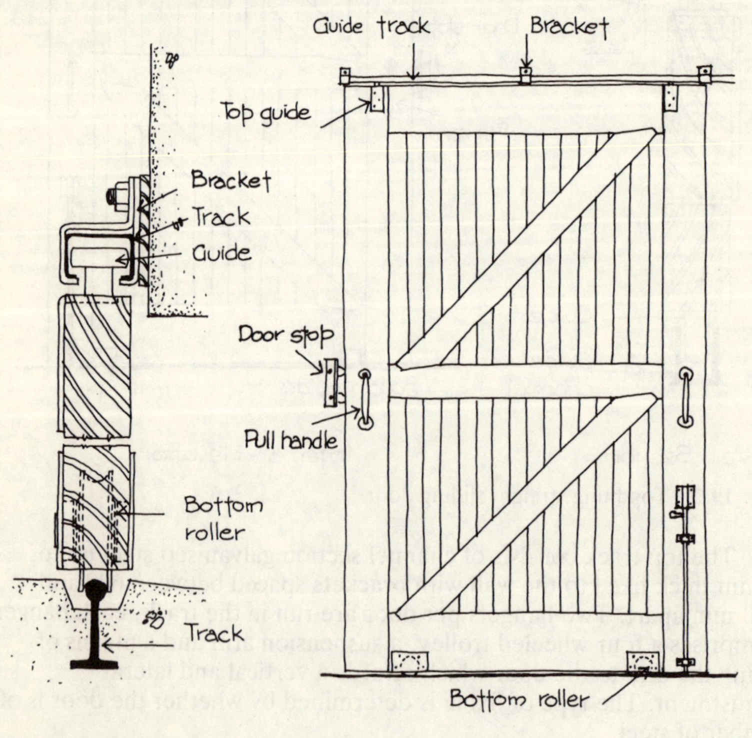

Fig. 19.3   Bottom roller straight sliding door

The bottom rollers, two per door, each consist of a single wheel, machined round the edge to suit the track in ball or roller bearings fixed to a steel plate casing which is let into the bottom of the door. At the top, two door guides are fixed to the door which run in a top track which is similar to the track for a top-hung door, only a lot lighter. These guides are either of solid steel or are a steel spindle with a horizontally operating nylon wheel.

When the door is fixed to slide along the outside of the wall the top track should be protected by a weatherhood.

19.3 Folding doors

These, like the sliding doors, can be either top hung or supported on bottom rollers. Tracks for each are much the same, the significant difference being that the point of support or suspension is through the pivot of a hinge screwed to adjacent door leaves (see Fig. 19.4).

The actual door is divided into leaves which are usually between 450 and 900 mm wide. Timber doors are connected by the top and bottom hinges of the door gear and at appropriate intermediate points by back-flap or butt hinges; steel doors, which can be much larger than timber, tend to have narrower leaves and continuous hinges plus a lattice work of steel bars on the inside face. The maximum practical size for timber doors is 3 m high and up to six leaves wide, whereas steel doors can be up to 6.5 m high and 21 m wide.

If the door is divided into three or five leaves, the end one at the leading edge is not held at the top and bottom and thus becomes a 'swinger'. This can be quite a useful asset in that it can be used as a normal door for access without having to move the rest, but it is recommended that a 'swinger' like this is opened first and fastened back to the adjoining leaf before the rest are folded back.

19.4 Draught exclusion

Both sliding and folding doors tend to be draughty because of the operating clearances which must be allowed. In many cases this is accepted because the use to which the building is to be put does not require warm, draught-free conditions. In other installations it is important both for the health and efficiency of the employees and to economise on heating fuels.

To meet this requirement, sliding doors can be arranged to slide past a timber post at each side of the opening, and the trailing edge of the door fitted with a projecting fillet which overlaps the post when the door is closed, thus sealing the jambs. The top and bottom edges can be sealed by a draught-excluding strip of the brush welt type which remains in contact with the lintel face or the paving as the door moves.

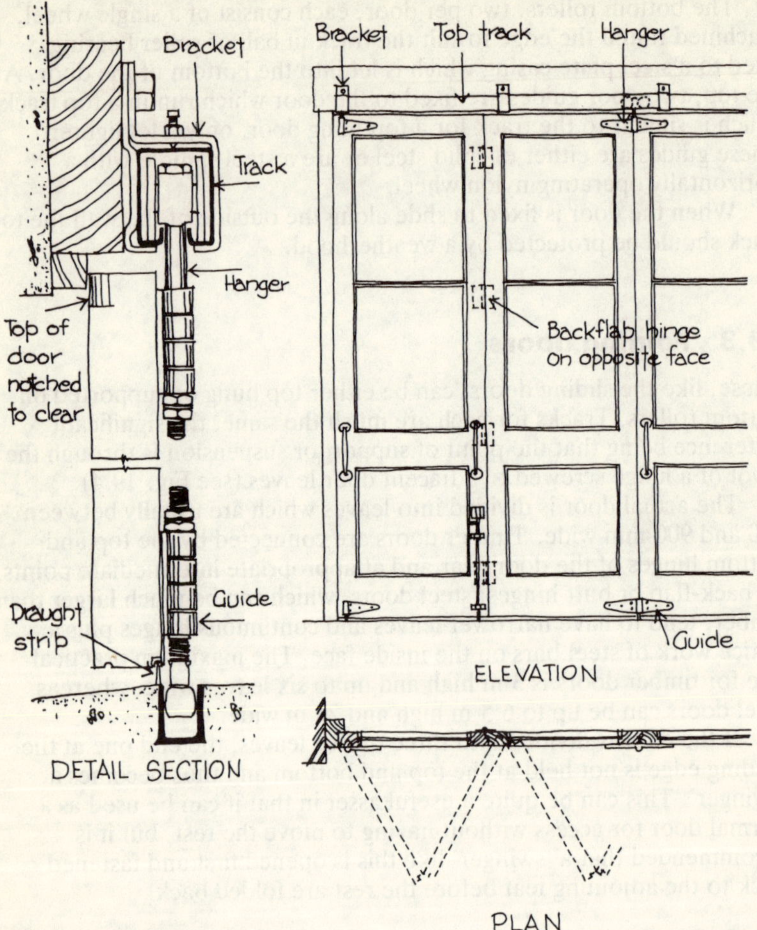

Fig. 19.4 Folding doors

Pairs of doors would have rebated meeting edges into which can be fitted a compressible draught-excluding strip. Folding doors can be similarly sealed at top and bottom with a brush welt, the outer edges of the door can fit into a rebated door post at each reveal and the edges of the leaves can also be rebated and a compressible sealing strip inserted.

19.5 Roller shutters and slideover doors

Roller shutter doors are composed of a shutter curtain made up from steel or aluminium interlocking laths which are rolled up onto a steel tube barrel across the top to open the door. At each side the laths are fitted into channel guides. The operation of small shutter doors is simply by pushing the door up, and the barrel is fitted with a self-coiling device to ensure the shutter curtain becomes correctly wound round the barrel. Large shutter doors are too heavy to operate in this manner and are fitted with bevel gears and a continuous chain drive which is manually operated to rotate the barrel and wind up the shutter curtain. Power operation is quite easy to fit to this type of door (see section 19.7).

Figure 19.5 shows a typical roller shutter installation and from this it can be seen that the system leaves the walls and floor completely

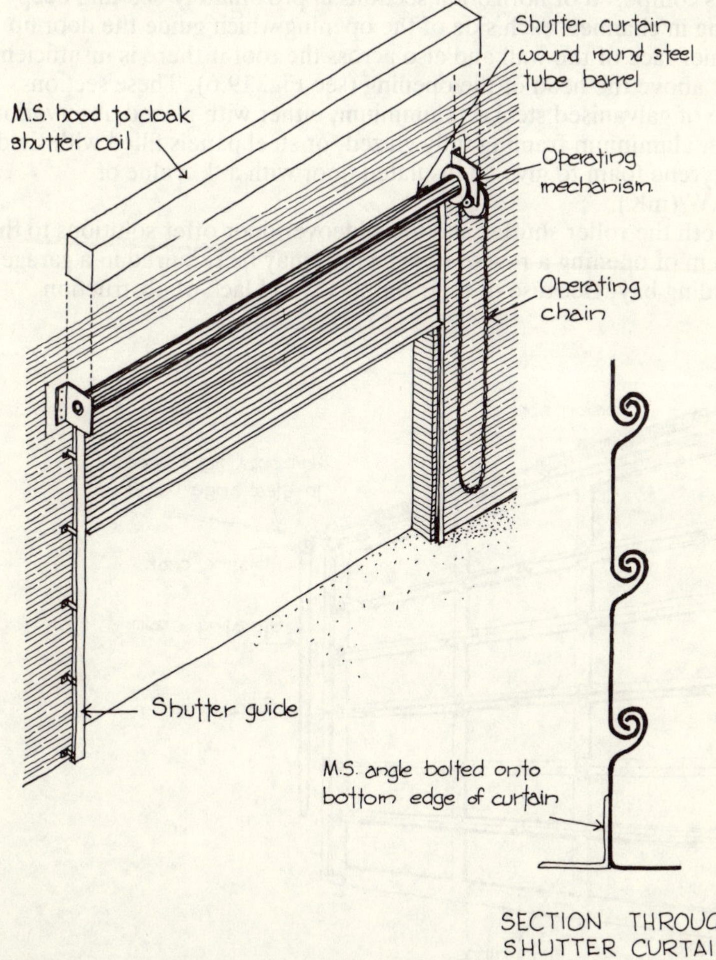

Fig. 19.5   Roller shutter door

unobstructed when the door is opened; however, it does place all the load of the door on the lintel. There is also a large box to accommodate across the door head in which the shutter is coiled. It is usual to restrict the height to a maximum of 7 m to keep the size of the shutter box reasonable and the width to a maximum of 12 m because of the lateral strength of the laths, and to reduce the overall weight of the door both for the convenience of opening and to avoid an excessive load on the lintel.

A drawback to the shutter door is that the laths, which are about 100 mm wide at their maximum, do not permit any reasonable glazing to be fitted into the door. This problem is overcome by the slideover door which is made up as a shutter door except that instead of laths the door is composed of horizontal sections approximately 600 mm deep running in channels each side of the opening which guide the door up the inner face of the wall and also across the roof if there is insufficient height above the head of the opening (see Fig. 19.6). These sections can be of galvanised steel or aluminium, either with glazed observation panels, aluminium framing fully glazed, or steel panels filled with rigid polystyrene foam to give an insulated door with a 'k' value of 0.036 W/(mK).

Both the roller shutter and the slideover door offer solutions to the problem of opening a row of doors which may be required in a garage or loading bay. Because of their simplicity and lack of obstruction

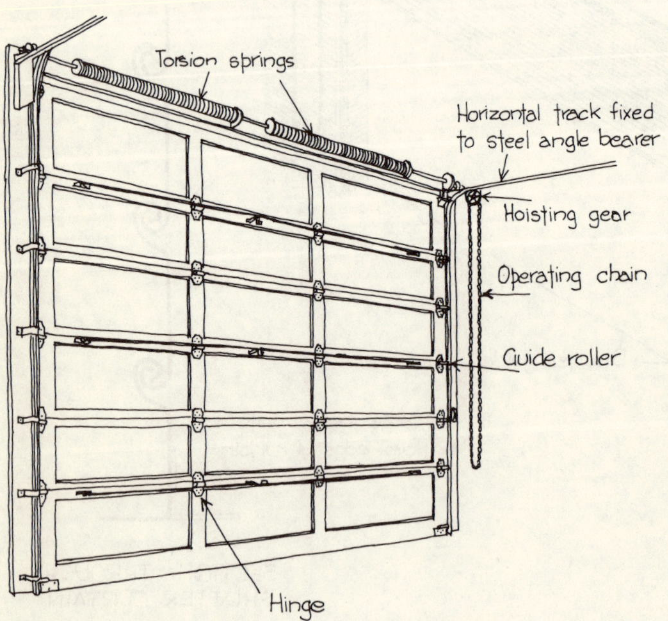

Fig. 19.6 Slideover door

they are ideal for this application and when opened will only leave a stanchion and two small channel section guides between each doorway.

19.6   Fire-resisting doors

Because of the clearances necessary to allow the opening of most of the doors mentioned in this chapter it is virtually impossible to render them fire resisting but as they are frequently a final exit door anyway they do not often require this property. There is one type which can be fire resisting and that is the roller shutter.

The roller shutter is required to be constructed in steel, usually galvanised for protection, to the specification laid down in the Fire Officers' Committee Rules, Section 1, Specification 3, and Sections 20 and 21 of the London Building Acts. The slats must be interlocking and of 18 gauge metal and fitted with special flame-proof end locks into an extra deep channel set into the brick reveal each side. At the bottom the shutter curtain is fitted with a double mild steel angle. With this construction fire resistance of two or four hours is obtainable.

In normal circumstances the shutter is operated either manually (if it is under 2400 mm wide and 2100 mm high) or by mechanical gear-operated hand wheels set into the adjoining wall up to a maximum size of 4250 mm wide × 3600 mm high. In the event of a fire a fusible link parts at 68°C, releasing a drop weight which automatically closes the door. Additional protection can be achieved by connecting the weight-retaining mechanism to the general alarm or smoke detection system so that the door closes even when the fire is some distance away.

Where, for a particular reason, a sliding or folding door system is preferred but protection in the event of a fire must be provided, a fire-resisting shutter can be installed between the jambs of the opening and the sliding or folding door fixed on the line of the inside face. In normal conditions the fire shutter stays up, but if the alarm is set off or the fusible link melts it drops and seals the opening whether the other doors are open or closed.

19.7   Door operation

All the doors described in this chapter have either manual or mechanical operation. Manual operation is restricted to the weight of door a person can move, which for sliding and folding doors can be more than for shutters or slideover doors.

Mechanical operation can either be a direct continuous chain loop system which is pulled or passed round a hand wheel or a system of bevel gears and rods, or it can be an electrically operated system whereby the door is opened or closed by an electric motor which is switched on by a variety of control methods and switched off by a limit switch at the end of the door travel.

The control methods used to switch on the electrical gear depend on the type of use but mainly comprise a push button or pull switch for the benefit of pedestrians. Inductive detectors set 50 mm under the road surface which respond to all metal vehicles or a photo electric beam across the approach to the door not only opens the door when the beam is interrupted but will hold the door open if a train of vehicles has to pass through.

An emergency manual release is usually available to allow the doors to be operated in the event of a power failure. This is a particularly important provision where the inability to open the door could trap someone in dangerous conditions such as a cold store, air lock, etc.

Chapter 20

Internal joinery and adhesives

20.1 Production management

The efficient management of the manufacture of joinery is basically a matter of matching demand and production capacity. Where a firm is making joinery items to order, the emphasis is on the control of the demand so as to ensure that the right quantity of work is put into the works at all times. There are many firms who produce standard joinery in quantities based on anticipated general demand and in their case the demand is fairly constant and production must be controlled to suit.

20.1.1 Purpose-made joinery production

Many buildings are 'one-off' jobs, individually designed and not be repeated. This does not mean that standardised mass-produced building elements or components cannot be used – the bricks will be the same as any other building and the windows may well be standard frames – but many of the parts and fittings will require special designs and manufacture.

This purpose-made work is frequently of timber and may be either a structural element in softwood such as a staircase, or a highly finished decorative item in exotic hardwoods, such as a bank counter.

It may be included in the general building contract as a single item to be made up in the contractor's own joinery shop, such as a decorative entrance canopy, or it may be a larger item requiring specialised working, such as veneered room panelling for which separate estimates are required. On the other hand it may be the subject of a separate contract for such items as moveable furniture. Whatever the contractual arrangement, the joinery shop management will only be concerned with programming the work into the production time allowed so as to remain within the estimated cost, not with numbers of unit, because only one is required.

20.1.2   Flow production of joinery

Purpose-made joinery is one of the range of production methods, the other is flow production in which a variety of standardised units is being produced at a constant rate. These units cover a wide range of goods from television cabinets to standard windows and account for a large proportion of the joinery output in this country.

The job of management in this case is to estimate the rate of demand per annum and adjust their production rate to suit. If the pattern of demand is variable, i.e. more garden seats sold in the summer than the winter, adequate storage facilities must be arranged to allow a constant rate of production.

20.1.3   Batch production of joinery

This falls between purpose-made and flow production; it is an intermittent flow production adjusted to suit a smaller demand which does not absorb the full annual production capacity. The system is that the joinery shop makes a batch of items of a standardised nature, i.e. staircases, these are then put into store and drawn upon as required; when the number in the store is reduced to a predetermined figure another batch is made and stocks replenished.

The problems involved with this method are concerned with deciding the size of each batch and when to make the next one. The second question is fairly easily decided from the rate of demand and the time taken to replenish the stocks. For instance, if the items are being drawn from stock at an average rate of five per week and it takes three weeks to produce a batch, a replenishment batch must be started when the stocks are down to 15 or perhaps 20 to allow for fluctuation. Starting too soon would lead to storage problems, and starting too late could mean running out of stock.

The first question, the size of a batch, is a matter for skilled judgement and management expertise. One method is to calculate the economic manufacturing quantity (EMQ), i.e. the batch size resulting in the lowest unit cost, taking into account such factors as costs of setting up the machinery, holding the stocks of the work in progress and also of the finished goods as well as the cost of the unit itself, the factory capacity and the demand for the product.

A formula to calculate the EMQ has been developed by the Furniture Development Council as follows:-

$$Q = \frac{200CsY}{IC}$$

where

Q = the minimum cost batch size
Cs = set-up cost of the batch
Y = annual production of the unit
I = cost of storage as a percentage of C
C = total cost per unit of manufacture

This, however, is subject to a certain criticism as it does not take into account the cost of work in progress which becomes increasingly important as the rate of demand rises.

Figure 20.1 shows, in graphical form, the effect on cost of varying the batch size away from the EMQ and it will be noticed that under-production is more uneconomic than over-production.

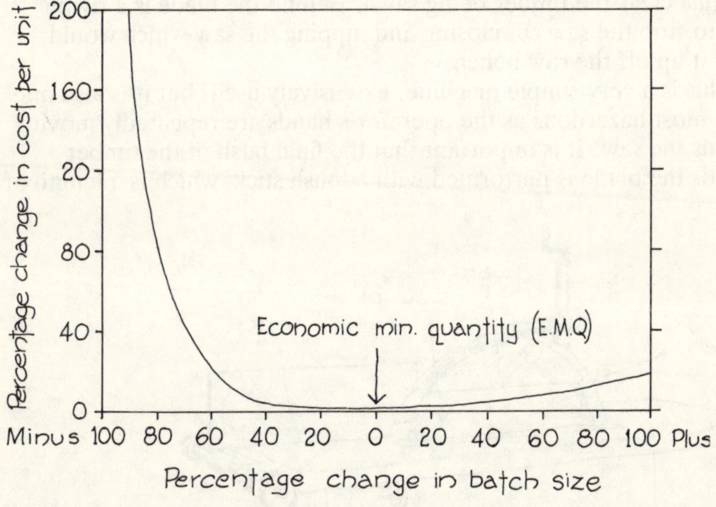

Fig. 20.1   Relationship between batch size and cost

20.2   Woodworking machinery

There is now a static wood-cutting machine to carry out all the basic operations normally carried out in the manufacture of timber; most joinery is designed so that it can be made with the minimum of hand work because of the cost that this entails. There is also a large family of portable power tools used by joiners to improve their efficiency, in many cases performing exactly the same operation as the static machine but neither so powerfully nor so accurately.

The machines mainly used in a joiner's shop are the circular saw, the cross-cut saw, the planer, the thicknesser, the morticer, the tenoner and the spindle moulder. Larger shops may have a band saw, dimension saw, vertical panel saw, various sanders and very large shops would include the power-fed machines which take a piece of sawn timber and produce the desired component in a single operation.

20.2.1   Circular rip saw

The circular saw bench is a basic machine in any woodworking shop since it performs the operation of ripping timber to size lengthwise which, without the machine, is a very laborious and time-consuming job.

Figure 20.2 shows a circular rip saw bench which consists of a table, a saw blade protruding through it and an adjustable fence against which the timber is pressed to guide it past the saw. Adjusting the fence produces the required size of timber. The saw blade is protected by a guard with an adjustable extension which must be set so as to just clear the timber being sawn. Behind the blade is a riving knife to stop the saw cut closing and nipping the saw which would throw it up off the saw bench.

This is a very simple machine, extensively used, but it is also one of the most hazardous as the operator's hands are repeatedly moving towards the saw. It is important that the final push of the timber towards the blade is performed with a 'push stick' which is a length of

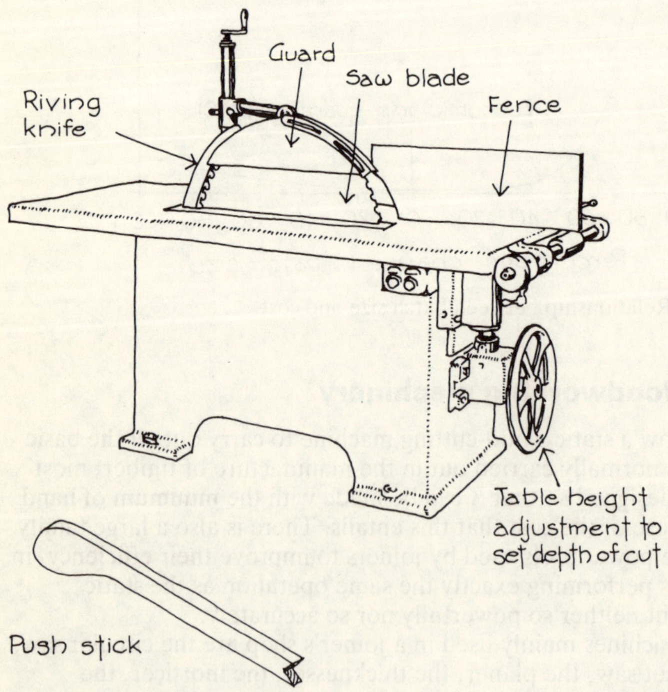

Fig. 20.2   Circular rip saw

small timber with a notched end (see Fig. 20.2) which prevents the operator's fingers being cut if the timber slips.

20.2.2 Cross-cut saw

The cross-cut saw is exactly opposite to the circular rip saw. It is designed to cut timber across its width, not along it, the timber remains still and the saw moves, and the saw and motor are above rather than below the work bench.

In operation, the timber to be cut is laid on the bench, as shown in Fig. 20.3, with the end against an adjustable stop if a batch is being prepared, and the saw pulled across.

The head is adjustable for angle so that mitres and complex bevels can be cut or a special trenching head can be fitted to cut a groove or housing across the face of the timber.

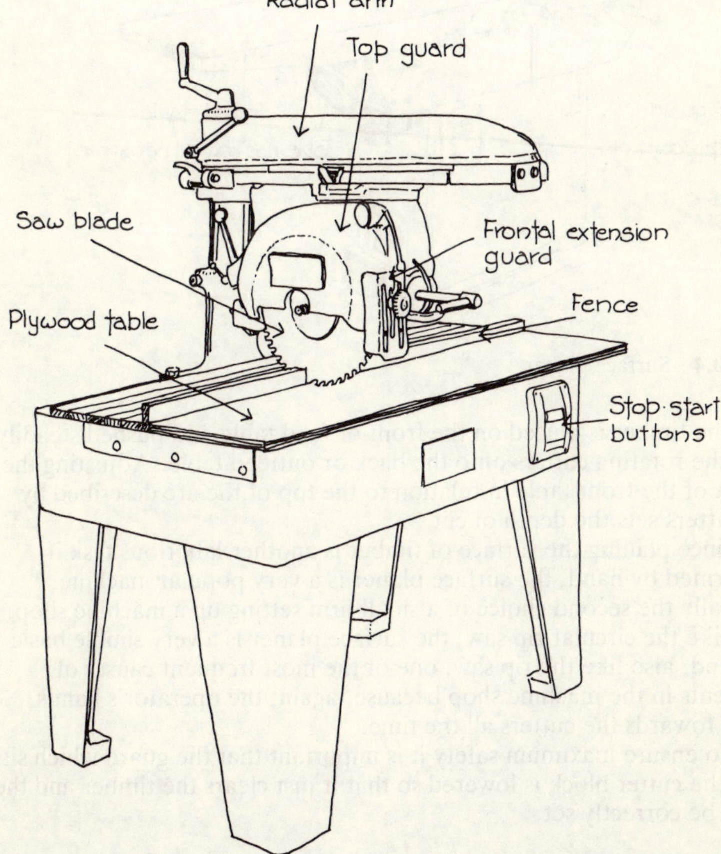

Fig. 20.3 Radial arm cross-cut saw

20.2.3 Surface planer

Once the timber has been cut to size by the rip saw, the next operation is to plane two adjacent sides to give a face side and a face edge from which the setting out dimensions on the timber are taken. This is the job of the surface planer, shown in Fig. 20.4, which consists of a work table in two parts, with a cutter block mounted on a horizontal spindle between them, and an adjustable fence to guide the timber. The cutter block is cylindrical and is fitted with two blades diametrically opposed so that the block is balanced.

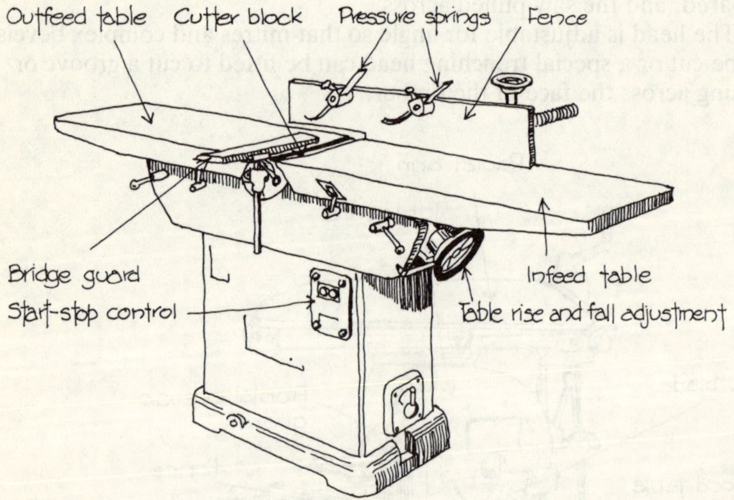

Fig. 20.4 Surface planer

The timber is placed on the front or feed table and pushed steadily over the rotating cutters onto the back or outfeed table. Adjusting the height of the front table in relation to the top of the arc described by the cutters sets the depth of cut.

Since planing the surface of timber is another laborious task if performed by hand, the surface planer is a very popular machine, generally the second choice of a small firm setting up a machine shop.

Like the circular rip saw, the surface planer is a very simple basic tool and, also like the rip saw, one of the most frequent causes of accidents in the machine shop because, again, the operator's hands move towards the cutters all the time.

To ensure maximum safety it is important that the guard which sits over the cutter block is lowered so that it just clears the timber and the fence be correctly set.

20.2.4 Thicknesser

Following planing the face side and edge, the next step is to plane the other two faces to give the required finished size of timber. This operation is carried out on a thicknesser which is similar to a surface planer except that the timber passes under the cutter block on a feed table set below the cutter arc at a distance equal to the specified thickness of the timber.

In this machine, the timber is taken past the cutter by a power feed arrangement, usually two rollers in contact with the upper surface of the timber, so that all the machinist has to do is to offer the timber into the machine and collect the finished work on the other side thus making it much less dangerous to use.

Because the thicknesser and the planer are very similar machines they are frequently combined into one piece of equipment so that the work is passed over the cutters to surface plane it and then under to thickness it (see Fig. 20.5).

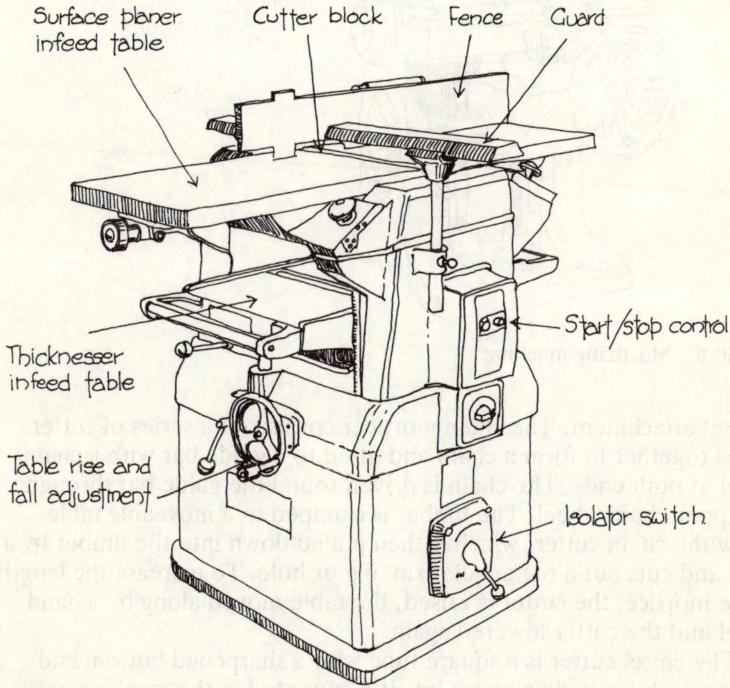

Fig. 20.5 Combined planer and thicknesser

20.2.5 Morticer

A number of machines have been developed to assist in the formation of joints and of these, the morticer and its complementary equipment, the tenoner, are the most common since the mortice and tenon joint is used more often than any other.

Figure 20.6 shows a morticing machine which has both a chain and

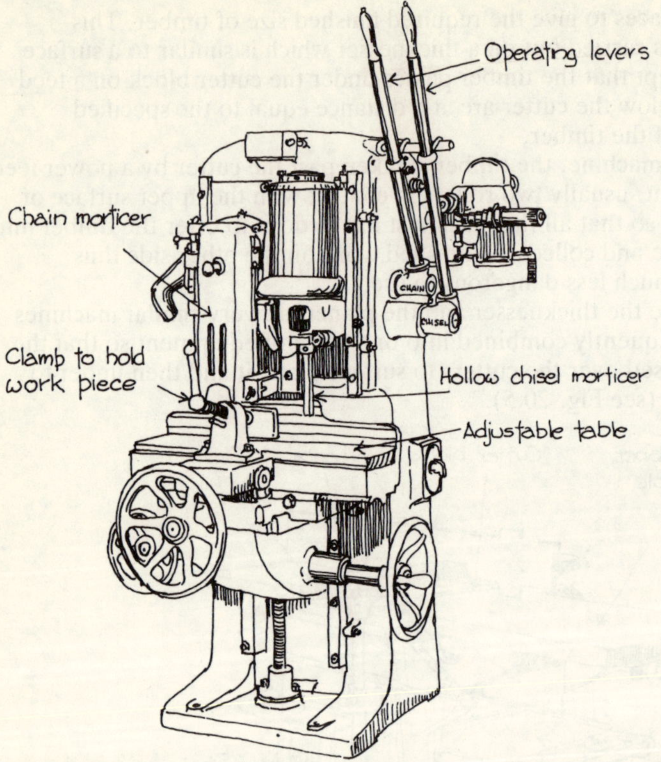

Fig. 20.6 Morticing machine

a chisel attachment. The chain morticer consists of a series of cutters linked together to form a chain and fitted to a guide bar with a chain wheel at both ends. The chain is driven round the guide bar through the upper chain wheel. The timber is clamped to a moveable table below the chain cutter, which is then pulled down into the timber by a lever and cuts out a rectangular sinking or hole. To increase the length of the mortice, the cutter is raised, the table moved along by a hand wheel and the cutter lowered again.

The chisel cutter is a square tube with a sharpened bottom end inside which is fitted an auger bit. It is mounted in the machine as shown, the timber is clamped to the table and when the cutter is lowered the auger drills a round hole and the chisel follows to make it into a square hole. By retracting the chisel, moving the table and lowering the chisel again the square hole can be converted to a rectangular mortice.

20.2.6   Tenoner

This is the machine which produces the other half of the mortice and tenon joint and is more complicated than any of the machines described so far. In operation the timber is clamped to a trolley running on a pair of guides and passes first between a pair of horizontally mounted cutter blocks which cut the tenon and then past vertically mounted cutter blocks which shape the shoulders. Finally it is pushed past a small circular saw which trims the tenon to length (see Fig. 20.7).

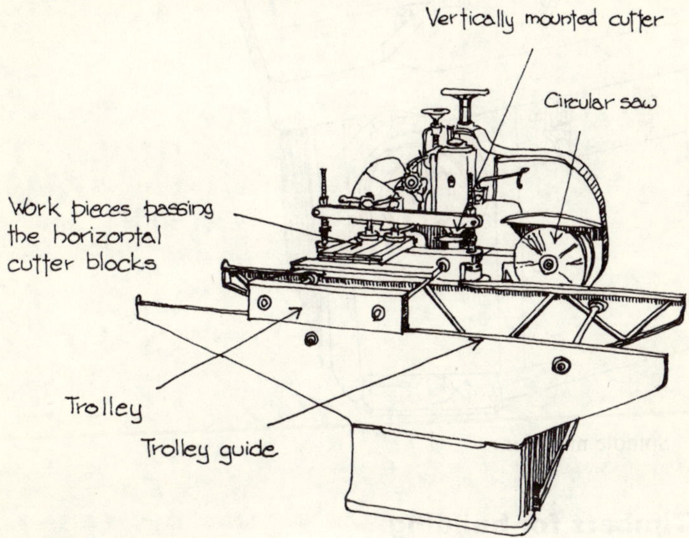

Fig. 20.7 Single end tenoner

With four cutter blocks to assemble and position and the cut-off saw to adjust, the tenoner is a machine which takes a long time to set up. For this reason it is only really economic where large quantities are required. For very large numbers a double-ended tenoner is available which is really two tenoning machines mounted on each side of the trolley guides, so that both ends of the timber are worked at the same time.

20.2.7 Spindle moulder

The principle of a spindle moulder is that of a vertical spindle which projects up through the worktable driven by a motor to rotate between 5000 and 15 000 revolutions per minute and to which a wide variety of cutters can be attached. The cutters are in pairs to obtain the correct balance and are ground to a desired shape so as to produce a moulding on timber being fed past them (see Fig. 20.8).

The spindle moulder is probably the most versatile of the basic woodworking machines in that, by the use of the correct cutter block and cutters, it can carry out a wide range of operations on both straight and curved timber.

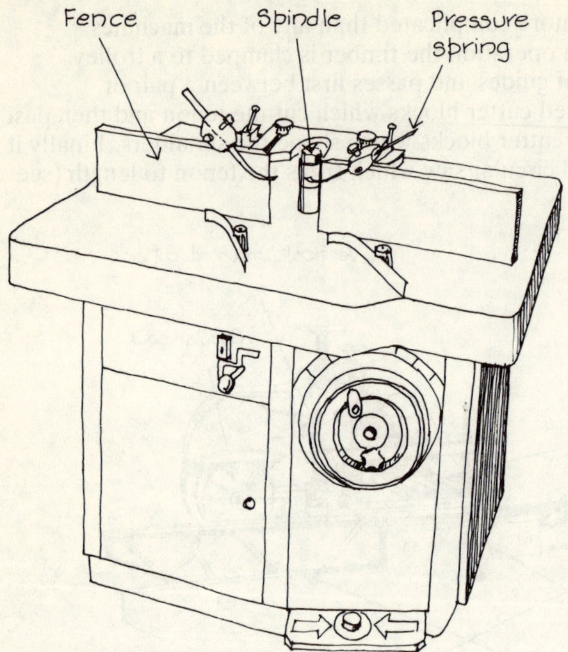

Fig. 20.8   Spindle moulder

20.3   Timbers for building

Timber is a very variable material. The same species of tree growing in differing areas or conditions will produce timber with dissimilar characteristics. To add to the confusion, there are several thousand varieties of tree each with a separate botanical name, but many varieties may be sold under the same trade name or the same variety sold under differing names related to the source of supply or port of shipment.

Furthermore, many names are confusing; Douglas fir, for instance, is also called British Columbian pine or Oregon pine; however, it is neither a fir nor a pine: Scots pine and Scots fir are both the same (in this case a pine), and there is no such timber species as 'deal' (it is actually a size of timber) but yet white, red and yellow deal are listed.

Botanically there are two types of tree, the angiosperms and the gymnosperms. The first are the deciduous, broad-leaved trees which mainly lose their leaves in the winter (an exception to this is holly), from which we obtain the hardwoods, and the second are the coniferous, needle-leaved trees which give us softwood.

The terms hardwood and softwood have nothing to do with the relative hardness or softness of the timbers; an extremely hard timber, Yew, has needle leaves which are not shed in winter and is thus a softwood whereas balsa wood, the softest timber of all, is actually classed as a hardwood.

British Standards 881 : 1974 589 : 1974 set out both the common name and the Latin name for each species. Use should be made of the recommended names, and if a particular tree or timber is required it should be specified by its Latin botanical name.

Table 20.1 lists some of the more commonly available softwoods and Table 20.2 lists hardwoods with their characteristics and use. Where the same timber is known by different names these are all shown together under the name which is most frequently used.

20.4 Adhesives

The adhesives used in the manufacture of glued laminated structures were mentioned in Chapter 10 and are exactly the same as used for general joinery purposes. There are three types of adhesive, all of which can be used for joinery purposes; they are animal glues, thermoplastic adhesives and synthetic resin thermosetting adhesives.

Animal glues derived from hides, skins and bones are now rarely used but casein glue, as mentioned in Chapter 10, which is obtained from sour milk, is used for internal joinery. It cannot be used externally since it loses its strength if wetted.

Thermoplastic adhesives are the polyvinyl acetate (PVA) range of adhesives. It is a white liquid of a creamy consistency which is transparent when set, easily applied, does not blunt tools and nor does it stain hardwood as casein glues can do. It is not suitable for situations where the glue line has to withstand a continuous stress. Shelf life (the time it can be stored) is at least six months and when used sets by loss of water.

The synthetic resins are urea-, phenol-, melanine-, and rescorcinol formaldehyde, all very strong and some are highly resistant to moisture. They set by the action of heat but the addition of a catalyst in the form of a hardener makes room temperature an adequate source of the required heat.

Urea formaldehyde is colourless, widely used in building work not exposed to the weather and in the manufacture of plywood and chipboard. It is classified MR (moisture resistant) which means it can withstand cold water only and is only moderately weather resistant.

Phenol formaldehyde is classed WBP (weather and boil proof) and supplied in the form of a film which requires applied heat to set (used for marine plywood manufacture) or as an adhesive which sets at room temperature for general joinery purposes.

Melamine formaldehyde is colourless like urea formaldehyde and supplied either as a film or in powder form to be mixed with water. Both the film and the solution need applied heat to set and the

Table 20.1 Common softwoods

Name	Average density (kg/m^3)	Characteristics and colour	Use
Douglas Fir Columbian Pine Oregon Pine	530	Straight grained and resilient. Easy to work. Strong and hard. Resists acids, decay and wear. Reddish-brown to pinkish-brown.	First-class joinery. Structural work. Glulam work. Plywood.
European Larch	600	Resinous. Straight grained. Very strong and durable. Reddish-brown.	Carpentry generally. Fencing, gates. Flooring. Railway sleepers.
European Spruce Whitewood Baltic whitewood Northern whitewood	430	Strong for its weight. Easily worked. Less durable than redwood. All pale yellow to pinkish-white.	Flooring. Blockboard cores. Internal carpentry. Temporary work.
Parana Pine	550	Even texture and straight grain. Unsuitable for exterior work. Brittle – not durable. Light to darkish-brown + some red.	Interior joinery. Cupboards. Stairs.
Pitch Pine	670	Highly resinous. Very durable, strong and heavy. Easy to work but resin clogs tools. Light red.	Shipbuilding. Heavy construction. Church work.

Redwood Scots Pine Red Pine Red deal Yellow deal	530	Straight grained. Easily worked. Very durable. Large knots can be troublesome. Pale reddish-brown.	Interior and exterior work. Beams, joists and rafters. Flooring. Window and doors. Fitments.
Western Hemlock Pacific Hemlock Brit. Columbian Hemlock	480	Uniform texture. Strong and durable. Straight grained and easily worked. Light brown.	Interior joinery. Built-in furniture. Timber buildings.
Western Red Cedar	380	Large clear boards available. Resists weather and insect attack. Not strong. Reddish-pink to dark brown.	Roof shingles. Cladding. Joinery. Decorative work.

Table 20.2 Common hardwoods

Name	Average density (kg/m^2)	Characteristics and colour	Use
African Mahogany	520 to 720	Cheapest of mahoganies. Difficult interlocked grain. Resistant to decay. Light brown to deep red.	Panelling. Furniture. Veneered work.
Afrormosia	690	Very durable. Fine grain. Stained by iron. Similar to teak, light brown.	High-class joinery.
Birch	670	Straight grained. Can be bent. Medium to fine texture. White to light brown.	Plywood.
Cuban Mahogany Spanish Mahogany Jamaican Mahogany	790	Beautifully figured. Hard. Medium to fine texture. All deep rich red or brown.	High-class joinery. Shopfitting. Furniture.
English Oak	720	Extremely hard, strong and tough. Susceptible to fungi attack. Discoloured by iron. Light yellowish-brown.	Every type of woodwork.
European Beech	720	Hard, close-grained. Fine texture. Durable. Reddish-yellow to light brown.	Furniture and handles. Block flooring. Parquet flooring. Veneers.
Gurjun	720	Dense. Very oily. Straight grained. Reddish-brown.	General construction.

Honduras Mahogany	540	Many grain figures available. Straight or interlocked grain. Excellent to work. Light reddish- to yellowish-brown.	
Iroko	670	Very hard, durable and strong. Well figured. Dulls cutter edges. Pale yellowish-brown to chocolate.	Internal, and external joinery. Doors. Windows. Laboratory benches, etc.
Jarrah	900	Hard, dense and heavy, Difficult to work. Resists fire. Deep rich red.	Flooring. Sleepers. Road blocks.
Sapele	640	Close textured. Much grain variation. Almost as hard as oak. Dark reddish-brown.	High class joinery. Shopfittings. Furniture. Veneered work.
Sycamore Plane	620	Close grained, even and fine. Strength equal to oak. Silky finish. Both white with tinge of yellow.	Flooring. Panelling. Fittings. Veneers and inlays.
Teak	720	Very strong and durable Resists water, insects and fire. Difficult to work. Golden brown.	High-class joinery. Doors. Bench tops. Draining boards.

resultant glue is classified as BR (boil resistant).

Rescorcinol formaldehyde sets at room temperature, after a hardener has been added, to produce an adhesive classified as WBP (water and boil proof). It forms an extremely strong and durable joint and has good gap filling qualities, i.e. is effective in a glue line up to 1.3 mm thick.

Part V

External works

Chapter 21

Rigid pavements

21.1 The functions of a road structure

The Department of Transport, Transport and Road Research
Laboratory with the Cement and Concrete Association have defined
the functions of a road structure in their joint publication, *A Guide to
Concrete Road Construction*, as follows:
1. To carry the wheel loads applied by traffic and to distribute the
   resultant loading over the soil beneath so that the stress induced in
   the soil does not exceed a safe value for the sub-grade.
2. To provide a running surface that will be safe and sufficiently strong
   to resist the effects of traffic and weather, particularly frost, with a
   minimum of maintenance over a long period.
3. To protect the underlying soil from an ingress of water and
   consequent variations in moisture content and thus assist in the
   maintenance, during the life of the road, of a uniform state of
   stability in the soil.

   In addition to these constructional characteristics, the road must
be designed to accommodate with safety the maximum volume of
traffic which can be anticipated travelling at the speeds officially
allowed, and it must also provide for safe use by pedestrians.

21.2 Terminology

The following are a selection of the terms used in connection with the
design of roads:

Pavement: The part of the road structure above the subgrade.

Sub-grade: The upper part of the soil, natural or constructed, which supports the loads transmitted by the overlying pavement.

Formation: The surface of the subgrade in its final shape after completion of the earthworks.

Sub-base: The material situated between the sub-grade and the slab or roadbase. It may be in one layer or in upper and lower sub-base layers.

Roadbase: The layer which provides the principal support to the surfacing. It may be a concrete slab or one or two layers of bituminous materials.

Slab: Plain or reinforced concrete which may be just the roadbase or may also provide the running surface.

Surfacing: The top layer providing the wearing coat. It may be applied bituminous material laid over a concrete or bituminous roadbase in one layer, or in two layers, as base course and wearing course.

Separation membrane: A layer of polythene, or similar sheeting, between a concrete slab and the subgrade.

21.3 Flexible and rigid pavements

The terms flexible and rigid pavements suggest that one form of road construction bends in use whilst the other stays flat. This is not the case, although a rigid pavement is more susceptible to cracking than a flexible one.

A flexible pavement is built up from a sub-base of compacted stone overlaid with a roadbase of graded crushed stone, single sized stone or granular material bound with bitumen or tar and surfaced with a wide range of material in either tar or bitumen binders. Where the binder is a hard bitumen the mixed material is referred to as asphalte, but where a softer binder of bitumen or tar is used it is referred to as coated macadam.

The sub-base for both flexible and rigid pavements is much the same, consisting of naturally occurring stone or crushed stone, in some cases stabilised by the addition of cement.

For a rigid pavement, the sub-base is overlaid with a concrete slab, which can either be in small panels and unreinforced, or in larger bays with a welded steel reinforcing mat, or longitudinal steel bars. If the concrete is to perform the function of both roadbase and running surface, the quality of the material must be controlled to give the necessary wearing resistance, and the top must be carefully tamped and worked to provide the correct surface texture. In some roads the concrete provides the roadbase only, and a wearing course is laid over it of bitumen or tar-bound granular material exactly as for a flexible road.

21.4 Design of concrete pavement bays

The objective in the layout of the joints in a concrete road is to control cracking of the slabs. Generally, this consideration requires joints of various types to be formed at intervals which produce smaller bay sizes than is possible to cast in one day's working. Indeed, as explained in section 21.5, the joints running the length of the road (longitudinal joints) may not be joints at all but induced cracks in a slab cast in one operation.

The length of a slab, i.e. the distance between transverse joints of any type, depends on whether the slab is reinforced or not, and if so, the weight of the reinforcing steel used. For unreinforced slabs, it has been found that a maximum length of about 5 m must be set if uncontrolled cracking is to be avoided. Reinforced slabs can be much longer according to the thickness of the slab and the weight of reinforcement as shown in Fig. 21.1.

Because of the differences in the loading patterns of a road in the longitudinal and transverse directions, it is recommended that the width is made less than the length. Normally a width of 4.5 m is chosen, but this may be increased to 4.65 m to match the required width of traffic lanes on a trunk road.

Thickness of slab not less than (m.m)	Weight of reinforcement fabric not less than (kg/m²)	Maximum spacing of transverse joints (m)
170	2·6	16·5
200	3·2	20
250	4·4	28

NOTE:
Every third joint to be an expansion joint.

Fig. 21.1 Spacing of transverse joints

Complete road pavements can be laid in excess of 4.65 m wide in one operation but if this is done longitudinal crack inducers and tie bars must be introduced to create longitudinal joints at less than the 4.65 m spacing.

Whilst it is obviously important to consider the maximum size of a slab, it is also necessary to control the minimum size. Narrow strips can occur at junctions and it has been found that in these positions slabs narrower than 1 m are liable to crack at closely spaced intervals.

Cracks can also be initiated by holes or recesses formed in the slab for gullies, manholes or road studs. Where gullies or manholes cannot be arranged in any other position, they should be set in their own small section of slab, completely isolated from the main slab by preformed expansion material and the omission of any tie bars. Road stud recesses should either be immediately beside a longitudinal joint or at least 150 mm away from it.

21.5 Joints in rigid pavements

Concrete, in common with all building materials, changes its size as temperature and humidity vary and in consequence expands, contracts and warps. If a concrete road pavement is laid in one continuous strip, this movement will give rise to high stresses which will lead to cracking. The purpose of joints is to control this cracking either by the deliberate introduction of planes of weakness along carefully selected lines which the crack follows or by providing gaps in the concrete slab which are filled with compressible material thus allowing for expansion.

Transverse joints (those running across the road) are of four types, contraction joints, expansion joints, warping joints and construction joints. Contraction joints (see Fig. 21.2) are not true joints but planes of weakness along which the crack is induced to occur. To form this joint, a triangular timber fillet is laid on the sub-base along the line of the joint before the concrete is placed, and after it has been laid a

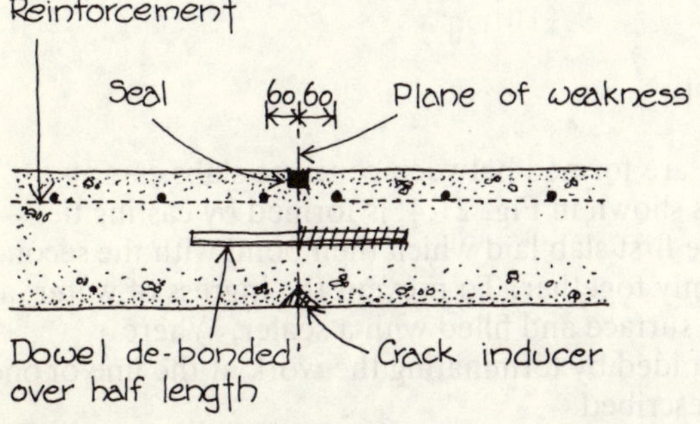

Fig. 21.2 Contraction joint

groove is formed in the top of the slab by vibrating a guillotine blade into the wet concrete. This is then removed, and after the concrete has set the groove is filled with a bituminous sealer to prevent the groove filling with stones and to stop water penetrating the crack below. An alternative method for the top groove is to vibrate a pre-formed compression seal into the surface and leave it there. In some cases, the concrete is left to set and the top groove cut in with a high-speed concrete saw.

Dowel bars are placed in the slab across the line of the crack to transfer the moving wheel load from one slab to the next, and half the bar is either sheathed in plastic, or coated with bitumen, to prevent it bonding to the slab, so that relative movement can occur between the slabs. These bars are either supported in cradles off the sub-base in advance of the concreting or else inserted into the wet concrete by a dowel bar placing machine.

A warping joint is simply a break in the concrete slab, tied together with bars which allows a small amount of angular movement to occur.

An expansion joint is formed by assembling the dowel bars and joint filler board into cradles fixed firmly to the sub-base. By this means, concreting can be carried out to both sides of the joint at the same time. To allow the necessary movement, one half of the dowel bar must be prevented from bonding, as for contraction joints, but in addition a waterproof cap containing a compressible filler must be fitted to provide room for the bar to move (see Fig. 21.3).

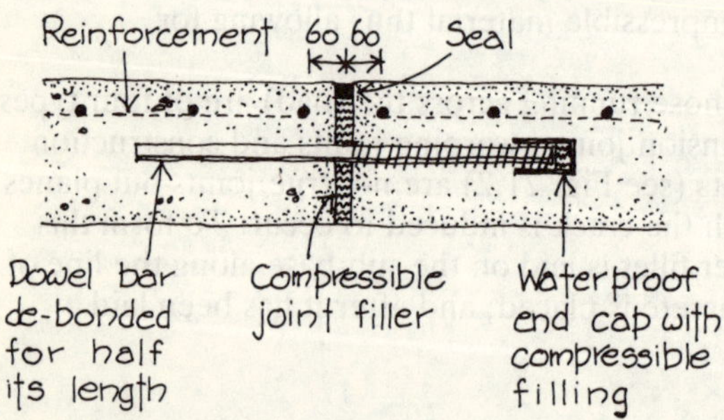

Fig. 21.3 Expansion joint

Construction joints are formed between concrete slabs cast at different times. This, as shown in Fig. 21.4, is formed by casting tie bars into the edge of the first slab laid which then bond with the second slab to hold the two firmly together. To prevent the ingress of water, a groove is formed in the surface and filled with a sealer. Where possible, this joint is avoided by terminating the work at the line of one of the previous joints described.

Longitudinal joints (those along the length of the road) are the

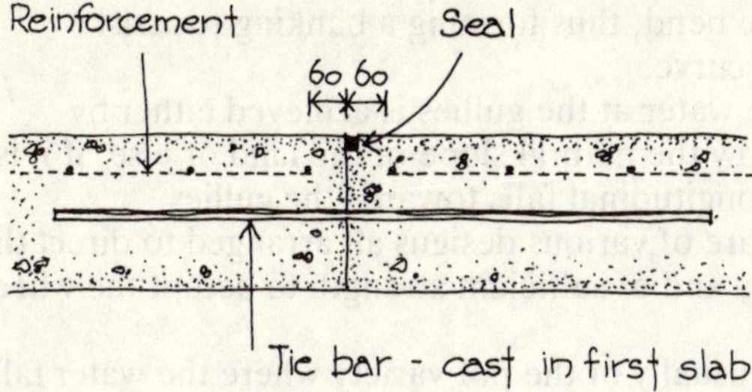

Fig. 21.4 Construction joint

same as contraction joints except that the steel bars are longer and are not de-bonded so that they act as ties across the line of the induced crack (see Fig. 21.5).

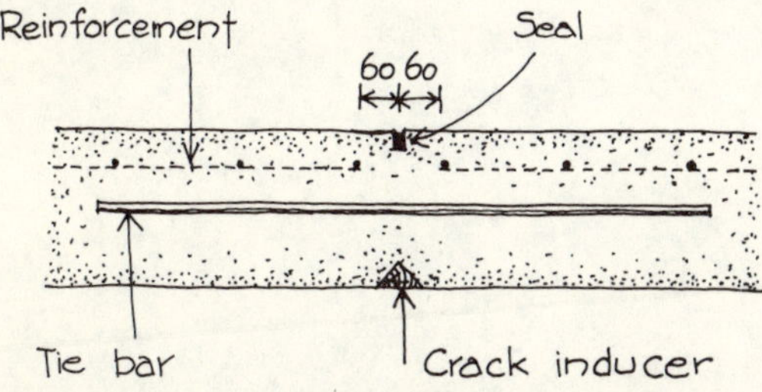

Fig. 21.5 Longitudinal joint

21.6 Road drainage

The efficient and fast removal of surface water on a road is essential for the safety of the road users. This removal process consists of three parts; the directing of the water off the running surface to the kerbs, the collection of the kerb edge water into gullies, and the disposal of the water in the gullies.

The directing of the water off the running surface is effected by inclining the surface an adequate fall. On a level, straight, fairly narrow road this is a simple matter of setting one edge of the road higher than the other to produce a cross fall. If the road is wide the necessary rate of fall can produce a large difference in kerb levels as well as a high concentration of water along one edge; in this case the surface is curved upwards with its high point along the centre line to form a cambered surface which sheds its water to both edges.

The rate of fall must be increased if the road is not level because then the longitudinal slope of the road causes the water to run diagonally to the edge, thus not clearing the surface so quickly.

On bends, a cross-fall must always be used with the high edge around the outside of the bend, thus forming a banking to assist vehicles to negotiate the curve.

The collection of the water at the gullies is achieved either by allowing the road to follow the natural slope of the land or else, if it is a level site, by forming longitudinal falls towards the gullies.

Road gully gratings are of various designs all arranged to direct the water down into the gully and of sufficient strength to accept the wheel loads of traffic.

The gullies used are usually of the pot variety where the water falls into a large chamber onto the side of which is fitted a trap through which the water drains away either to a surface water sewer or, if such is not available, to soakaways located away from the road pavements.

Bibliography

Royal Institute of British Architects, *Architectural Practice and Management*. RIBA Publications Ltd

House Information Services Ltd, *House's Guide to the Construction Industry*

Building Regulations 1976 and Amendments. HMSO

Royal Institute of British Architects, *CI/SfB Construction Indexing Manual*. RIBA Publications Ltd

Dennis H. Edmeads, *The Construction Site*. The Estates Gazette Ltd

J. T. Bowyer, *Small Works Supervision*. The Architectural Press Ltd

Mitchells Building Construction Series B. T. Batsford Ltd

M. J. Tomlinson, *Foundation Design and Construction*. Pitman

R. F. Craig, *Soil Mechanics*. Van Noastrand Reinhold

V. C. Launder, *Foundations*. The Macmillan Press Ltd

R. Chudley, *Construction Technology* Vol. 4. Longman

Allan Hodgkinson, *A.J. Handbook of Building Structure*. The Architectural Press Ltd

G. Barnbrook, *House Foundations for the Builder and Building Designer*. Cement and Concrete Association

Thomas Whitaker, *The Design of Piled Foundations*. Pergamon Press

Ronald E. Brand, *Falsework and Access Scaffold in Tubular Steel* McGraw-Hill

W. Thorley, *Design of Loadbearing Brickwork in SI and Imperial Units*. Heinemann

A. Gilmour Brown & R. Evans, *Production Planning and Control in the Furniture Industry*. Furniture Development Council

R. Bayliss, *Carpentry and Joinery*. Hutchinson Educational

Frank Hilton, *Advanced Carpentry and Joinery*. Longman

Dept. of Transport & Cement & Concrete Association, *A Guide to Concrete Road Construction*. HMSO.

Index